METHODS IN MOLECULAR BIOLOGY™

For further volumes:
http://www.springer.com/series/7651

Series Editor
John M. Walker
School of Life Sciences
University of Hertfordshire
Hatfield, Hertfordshire, AL10 9AB, UK

For further volumes:
http://www.springer.com/series/7651

Plant Signalling Networks

Methods and Protocols

Edited by

Zhi-Yong Wang

Department of Plant Biology, Carnegie Institution for Science, Stanford, CA, USA

Zhenbiao Yang

Department of Botany and Plant Science, University of California, Riverside, CA, USA

Editors
Zhi-Yong Wang
Department of Plant Biology
Carnegie Institution for Science
Stanford, CA, USA

Zhenbiao Yang
Department of Botany and Plant Science
University of California
Riverside, CA, USA

ISSN 1064-3745 e-ISSN 1940-6029
ISBN 978-1-61779-808-5 e-ISBN 978-1-61779-809-2
DOI 10.1007/978-1-61779-809-2
Springer New York Dordrecht Heidelberg London

Library of Congress Control Number: 2012936833

Printed on acid-free paper

Humana Press is part of Springer Science+Business Media (www.springer.com)

Preface

Signal transduction is the fundamental mechanism for regulation of cellular activities by environmental cues and regulatory signals. Signal transduction is particularly important for plants, whose survival requires proper physiological and developmental responses to the environmental changes. Genome sequencing has revealed expansion of gene families encoding signal transduction proteins in plants compared to animals. Genetic studies in the last two decades have identified receptors and key signal transduction components of many signaling pathways in plants, mostly in the model system Arabidopsis. However, plant signaling systems are complex and require diverse approaches and techniques to dissect. Conceptually, signal transduction involves signal perception by receptors and activation of receptor activity, intracellular signal relay, which often involves protein–protein interaction and posttranslational protein modifications such as phosphorylation, glycosylation, ubiquitination, and oxidation, and regulation of gene expression. Much progress has been made recently in the plant signal transduction research field thanks to the development of diverse techniques, including proteomics and mass spectrometry methods for studying protein modification, biochemical and cell biological tools for studying protein–protein interactions, genomic techniques for dissecting protein–DNA interaction and transcription networks, and computation methods that integrate molecular networks into plant developmental processes. *Plant Signaling Networks* describes many of these advanced research methods.

Chapters 1–3 describe mass spectrometry methods for studying protein phosphorylation and glycosylation. One of the most important classes of plant receptors is the receptor-like kinases localized on the cell surface. Methods for analysis of receptor kinase phosphorylation using mass spectrometry are provided in Chap. 1. These methods have yielded insights into the molecular details of receptor kinase activation by autophosphorylation and transphosphorylation in receptor kinase complexes. Chapter 2 describes quantitative measurement of protein phosphorylation in complex samples, which is useful in identifying phosphorylated components in signal transduction pathways. Chapter 3 describes enrichment and mass spectrometry analysis of O-GlcNAc modification of proteins, which is an important protein modification for signaling. Chapters 4–6 describe advanced two-dimensional electrophoresis methods for quantitative proteomic analysis of proteins localized on the plasma membrane, or modified by phosphorylation or redox.

Genetic approaches are powerful for identifying essential signaling components, but have limitations due to genetic redundancy. Several elaborate strategies have been shown to be effective in overcoming genetic redundancy. Chapters 7 and 8 describe chemical genetics—use of small molecule chemicals, to dissect signaling pathways, and Chapter 9 describes an improved tool for generating overexpression mutants.

G-proteins are important components of many signal transduction pathways. Chapters 10 and 11 describe biochemical and cell biological methods for analyzing G-protein activation. Ubiquitination is another universal mechanism used widely in all cellular regulatory processes. Specific interactions between E3 ubiquitin ligases and substrate proteins are key for regulating degradation/accumulation of signaling proteins. Chapters 12 and 13 describe in vivo and in vitro methods for analyzing E3–substrate interaction and ubiquitination.

Most signal transduction pathways regulate development and physiology by controlling gene expression. Identification of all target genes of a signaling pathway is key for understanding not only the functions of the pathway but also the regulatory network that integrates multiple pathways. Chapter 14 describes the use of chromatin immunoprecipitation followed by microarray (ChIP-chip) or high-throughput sequencing (ChIP-Seq) for identifying target genes of transcription factors in both *Arabidopsis* and rice. Quantitative analysis of gene expression as output of signal transduction provides effective assays for functions of signaling components. Chapter 15 describes a smart pooling approach that improves the efficiency of RNA profiling experiments. Chapter 16 describes the powerful cell-based transient gene expression assay for testing functions of and delineating relationships among signaling components. Chapter 17 describes a method for profiling un-capped RNA, which can reveal posttranscriptional regulation of RNA abundance. Finally, Chap. 18 provides brief account of recently developed imaging and computation methods for analyzing both local and global patterns of gene expression and growth in *Arabidopsis* shoot apical meristems (SAMs).

Plant Signalling Networks provides detailed protocols for a wide range of research approaches including genetics, proteomics, biochemical, cell biological, and computational approaches. These are powerful methods for understanding various aspects of signaling networks in plants. We hope that this timely overview of diverse approaches for studying signal transduction systems provides a guide for researchers to gain comprehensive understanding of complex signaling networks in plants.

Stanford, CA, USA — Zhi-Yong Wang
Riverside, CA, USA — Zhenbiao Yang

Contents

Contributors

DOMINIQUE AUDENAERT • *Department of Plant Systems Biology, VIB, Ghent, Belgium; Department of Plant Biotechnology and Bioinformatics, Ghent University, Ghent, Belgium*

TOM BEECKMAN • *Department of Plant Systems Biology, VIB, Ghent, Belgium; Department of Plant Biotechnology and Bioinformatics, Ghent University, Ghent, Belgium*

ANGELA BRUEX • *Department of Molecular, Cellular, and Developmental Biology, University of Michigan, Ann Arbor, MI, USA*

SHUOLEI BU • *Department of Plant Biology, Carnegie Institution for Science, Stanford, CA, USA*

ALMA L. BURLINGAME • *Department of Pharmaceutical Chemistry, University of California, San Francisco, CA, USA*

ANIRBAN CHAKRABORTY • *Department of Electrical Engineering, University of California, Riverside, CA, USA*

ROBERT J. CHALKLEY • *Department of Pharmaceutical Chemistry, University of California, San Francisco, CA, USA*

AMIT ROY CHOWDHURY • *Department of Electrical Engineering, University of California, Riverside, CA, USA*

STEVEN D. CLOUSE • *Department of Horticultural Science, North Carolina State University, Raleigh, NC, USA*

MIRELA-CORINA CODREANU • *Department of Plant Systems Biology, VIB, Ghent, Belgium; Department of Plant Biotechnology and Bioinformatics, Ghent University, Ghent, Belgium*

ZHIPING DENG • *Department of Plant Biology, Carnegie Institution for Science, Stanford, CA, USA; Institute of Virology and Biotechnology, Zhejiang Academy of Agricultural Science, Hangzhou, China*

YING FU • *State Key Laboratory of Plant Physiology and Biochemistry, Department of Plant Sciences, College of Biological Sciences, China Agricultural University, Beijing, China*

MICHAEL B. GOSHE • *Department of Molecular and Structural Biochemistry, North Carolina State University, Raleigh, NC, USA*

XIAOPING GOU • *School of life sciences, Lanzhou University, Lanzhou, China*

YULING JIAO • *State Key Laboratory of Plant Genomics, Institute of Genetics and Developmental Biology, Chinese Academy of Sciences, Chaoyang, Beijing, China*

RAGHUNANDAN M. KAINKARYAM • *Department of Chemical Engineering, University of Michigan, Ann Arbor, MI, USA; Procter & Gamble Company, Cincinnati, OH, USA*

JIA LI • *School of life sciences, Lanzhou University, Lanzhou, China*

NING LI • *Division of life science, The Hong Kong University of Science and Technology, Clear Water Bay, Hong Kong, SAR, China*

Lijing Liu • *State Key Laboratory of Plant Genomics, National Center for Plant Gene Research, Institute of Genetics and Developmental Biology, Chinese Academy of Sciences, Beijing, China*
Min Liu • *Department of Electrical Engineering, University of California, Riverside, CA, USA*
Srijeet K. Mitra • *Department of Horticultural Science, North Carolina State University, Raleigh, NC, USA*
Katya Mkrtchyan • *Department of Electrical Engineering, University of California, Riverside, CA, USA*
Long Nguyen • *VIB Compound Screening Facility, Ghent, Belgium*
Yajie Niu • *Department of Molecular Biology and Center for Computational and Integrative Biology, Massachusetts General Hospital, Boston, MA, USA; Department of Genetics, Harvard Medical School, Boston, MA, USA*
G. Venugopala Reddy • *Department of Botany and Plant Sciences, and Center for Plant Cell Biology, Institute of Integrative Genome Biology, University of California, Riverside, CA, USA*
José Luis Riechmann • *Division of Biology 156-29, California Institute of Technology, Pasadena, CA, USA; Center for Research in Agricultural Genomics (CRAG), Barcelona, Spain; Institució Catalana de Recerca i Estudis Avançats (ICREA), Barcelona, Spain*
Eugenia Russinova • *Department of Plant Systems Biology, VIB, Ghent, Belgium; Department of Plant Biotechnology and Bioinformatics, Ghent University, Ghent, Belgium*
John Schiefelbein • *Department of Molecular, Cellular, and Developmental Biology, University of Michigan, Ann Arbor, MI, USA*
Jen Sheen • *Department of Molecular Biology and Center for Computational and Integrative Biology, Massachusetts General Hospital, Boston, MA, USA; Department of Genetics, Harvard Medical School, Boston, MA, USA*
Yu Sun • *Department of Plant Biology, Carnegie Institution for Science, Stanford, CA, USA*
Wenqiang Tang • *Department of Plant Biology, Carnegie Institution for Science, 260 Panama Street, Stanford, CA, USA; Institute of Molecular Cell Biology, College of Life Science, Hebei Normal University, Shijiazhuang, Hebei, China*
Moses Tataw • *Department of Electrical Engineering, University of California, Riverside, CA, USA*
Hai Wang • *Department of Biology, Hong Kong Baptist University, Kowloon, Hong Kong*
Shengbing Wang • *Department of Medicine, John Hopkins University, Baltimore, MD, USA*
Zhi-Yong Wang • *Department of Plant Biology, Carnegie Institution for Science, Stanford, CA, USA*
Peter J. Woolf • *Department of Chemical Engineering, University of Michigan, Ann Arbor, MI, USA*
Yiji Xia • *Department of Biology, Hong Kong Baptist University, Kowloon, Hong Kong*

QI XIE • *State Key Laboratory of Plant Genomics, National Center for Plant Gene Research, Institute of Genetics and Developmental Biology, Chinese Academy of Sciences, Beijing, China*

SHOU-LING XU • *Department of Plant Biology, Carnegie Institution for Science, Stanford, CA, USA*

TONGDA XU • *Temasek Life Sciences Laboratory, National University of Singapore, Singapore, Singapore*

RAM KISHOR YADAV • *Department of Botany and Plant Sciences, and Center for Plant Cell Biology, Institute of Integrative Genome Biology, University of California, Riverside, CA, USA*

QINGZHEN ZHAO • *State Key Laboratory of Plant Genomics, National Center for Plant Gene Research, Institute of Genetics and Developmental Biology, Chinese Academy of Sciences, Beijing, China*

YANG ZHAO • *Institute of Plant Physiology and Ecology, Shanghai Institutes for Biological Science, Chinese Academy of Sciences, Shanghai, China*

JIA-YING ZHU • *Department of Plant Biology, Carnegie Institution for Science, Stanford, CA, USA*

LEI ZHU • *State Key Laboratory of Plant Physiology and Biochemistry, Department of Plant Sciences, College of Biological Sciences, China Agricultural University, Beijing, China*

Chapter 1

Experimental Analysis of Receptor Kinase Phosphorylation

Srijeet K. Mitra, Michael B. Goshe, and Steven D. Clouse

Abstract

Ligand binding by the extracellular domain of receptor kinases leads to phosphorylation and activation of the cytoplasmic domain of these important membrane-bound signaling proteins. To thoroughly characterize receptor kinase function, it is essential to identify specific phosphorylation sites by mass spectrometry. In this chapter, we summarize an efficient protein purification and modification protocol to prepare receptor kinases for liquid chromatography/tandem mass spectrometry analysis. Both recombinant receptor kinase cytoplasmic domains expressed in bacteria and full-length receptor kinase proteins expressed in living plant tissue are considered, and multiple methods of mass spectrometry are described that allow optimal identification of phosphorylated peptides of both in vitro- and in vivo-derived samples.

Key words: Receptor kinase, Phosphorylation, Phosphopeptide, Liquid chromatography/mass spectrometry, Immobilized metal ion affinity chromatography, *Arabidopsis*

1. Introduction

Reversible protein phosphorylation on specific Ser, Thr, or Tyr residues is a key component of many eukaryotic signal transduction pathways, resulting in altered protein function and turnover, changes in protein cellular localization, and modified protein–protein interactions. Receptor kinases play a pivotal role in many such signaling pathways. The mechanism of action of numerous plant and animal receptor kinases has been well characterized and generally involves recognition of a specific ligand by the extracellular domain, which mediates oligomerization of the receptor followed by phosphorylation and activation of the intracellular kinase domain (1–4). Such kinase activation allows recognition and phosphorylation of downstream substrates, leading ultimately to alterations in gene expression and corresponding changes in cellular physiology. To thoroughly characterize receptor kinase function, it is essential to understand the role of ligand-dependent

Zhi-Yong Wang and Zhenbiao Yang (eds.), *Plant Signalling Networks: Methods and Protocols*, Methods in Molecular Biology, vol. 876, DOI 10.1007/978-1-61779-809-2_1,

cytoplasmic domain phosphorylation, including identification of specific phosphorylation sites and characterization of their functional significance in the signaling pathway.

We are using genetics, kinase biochemistry, proteomics, and phosphoprotein analysis with liquid chromatography/tandem mass spectrometry (LC/MS/MS) to generate a phosphorylation site database for the large family of leucine-rich repeat receptor-like kinases (LRR RLKs) in *Arabidopsis thaliana* that have functional roles in the regulation of plant growth, morphogenesis, disease resistance, and responses to environmental stress signals (5). As a resource for this study, we have cloned over 200 LRR RLKs in various bacterial and plant expression vectors (6). LRR RLK cytoplasmic domains, including the juxtamembrane region, the catalytic kinase domain, and the short C-terminal domain, are expressed as recombinant proteins in *Escherichia coli* with a small N-terminal epitope tag (e.g., FLAG- or His-tag). After protein purification and autophosphorylation in the presence of ATP, in vitro phosphorylation sites can be determined by LC/MS/MS analysis. While we have found that in vitro LRR RLK phosphorylation sites can be highly predictive of in vivo phosphorylation, for a full functional characterization it is also essential to study the dynamic, ligand-dependent changes in phosphorylation of specific LRR RLK residues in planta. For these studies, epitope-tagged full-length LRR RLKs are expressed in *Arabidopsis* plants and a variety of LC/MS/MS procedures are used to examine in vivo phosphorylation sites for LRR RLKs immunoprecipitated from purified membrane fractions.

The BRASSINOSTEROID INSENSITIVE 1 (BRI1) and BRI1-ASSOCIATED RECEPTOR KINASE (BAK1), both critically involved in signal transduction for the essential plant steroid hormone brassinolide, have been extensively studied (7–11) and share some mechanistic similarities to animal receptor kinase function (12, 13). The methods described below were used in an exhaustive analysis of both in vitro and in vivo phosphorylation sites of BRI1 and BAK1. Immunoprecipitation of BRI1-Flag and BAK1-GFP from solubilized microsomal fractions isolated from brassinolide-treated transgenic *Arabidopsis* plants, followed by multiple LC/MS/MS approaches, resulted in the identification of 12 BRI1 phosphorylation sites and an additional 5 in BAK1 (12, 14, 15). A similar analysis of bacterial-expressed recombinant BRI1 and BAK1 cytoplasmic domains allowed comparative analysis of in vitro autophosphorylation sites as well as transphosphorylation sites between the two receptor kinases (10, 12). Methods were optimized and standardized for high-throughput analysis and have been applied to the identification of phosphorylation sites in over 100 LRR RLKs (SK Mitra, MB Goshe, SD Clouse, manuscript in preparation). While the protocols below mention specific epitope tags and defined LC/MS/MS approaches, it should be reasonably straightforward to modify the approach to accommodate other receptor kinase constructs and different LC/MS/MS instrumentation.

2. Materials

2.1. Recombinant Protein Expression, Extraction, and Purification Using Ni-NTA Beads

1. 250-ml Erlenmeyer flasks, LB medium, and shaking incubator.
2. Variable speed microcentrifuge.
3. Lysis buffer: 6 M guanidium hydrochloride, 300 mM NaCl, 1 mM phenylmethanesulfonyl fluoride (PMSF), and 20 mM imidazole.
4. Sonication probe (sonic dismembrator model F60, Thermo Fisher Scientific, Pittsburgh, PA).
5. Equilibration buffer: 50 mM Na_2HPO_4 (pH 8.0) and 300 mM NaCl.
6. Wash buffer: 50 mM Na_2HPO_4 (pH 8.0), 300 mM NaCl, and 100 mM imidazole.
7. Elution buffer: 20 mM Tris–HCl (pH 8.0), 100 mM NaCl, and 250 mM imidazole.
8. Agarose Ni-NTA beads (Qiagen, Valencia, CA).

2.2. In Vitro Kinase Reaction Using Unlabeled ATP

1. 1× Kinase reaction buffer: 50 mM HEPES (pH 7.4), 10 mM $MgCl_2$, 10 mM $MnCl_2$, 1 mM dithiothreitol (DTT), and 10 μM ATP.
2. Water bath.

2.3. In-Solution Tryptic Digestion

1. Sequencing grade trypsin (Promega, Madison, WI, no.: V5111).
2. Ammonium bicarbonate, acetonitrile, Tris–(2-carboxyethyl) phosphine hydrochloride (TCEP·HCl), and iodoacetamide.
3. Variable speed microcentrifuge and vacuum centrifuge.
4. Chloroform.

2.4. Purification of Microsomal Membrane Fractions from Plant Tissues

1. *Arabidopsis thaliana* (ecotype Columbia) seeds can be obtained from the Arabidopsis Biological Resource Center, Ohio State University, Columbus, OH.
2. Knife Blender [Oster (Shelton, CT) mini blender, Model Galaxie].
3. Centrifuge capable of 6,000 × *g* at 4°C.
4. Ultracentrifuge capable of 100,000 × *g* at 4°C.
5. Lysis buffer: 20 mM Tris–HCl, pH 8.8, 150 mM NaCl, 1 mM EDTA, 20% glycerol, 1 mM PMSF, 20 mM sodium fluoride, 50 nM Microcystin (Calbiochem, Gibbstown, NJ), one Protease inhibitor cocktail tablet (Roche Diagnostics, Indianapolis, IN) per 50 ml extraction buffer.

2.5. Membrane Protein Extraction

1. Sonication probe (sonic dismembrator model F60, Thermo Fisher Scientific).
2. Detergent compatible Bradford reagent kit (Biorad, Hercules, CA).
3. Polypropylene culture tubes.
4. Resuspension buffer: 10 mM Tris–HCl, pH 7.3, 150 mM NaCl, 1 mM EDTA, 10% glycerol, 1 mM PMSF, 20 mM sodium fluoride, 50 nM Microcystin, one Protease inhibitor cocktail tablet per 50 ml resuspension buffer, and 1% Triton X-100.

2.6. Immunoprecipitation

1. Dilution buffer: 10 mM Tris–HCl (pH 7.3), 150 mM NaCl, 1 mM EDTA, 10% glycerol, 1 mM PMSF, 20 mM sodium fluoride, 50 nM Microcystin, one Protease inhibitor cocktail tablet per 50 ml dilution buffer.
2. Anti-FLAG M_2 beads (Sigma, St. Louis, MO).
3. Wash buffer: 50 mM Tris–HCl (pH 7.3), 150 mM NaCl, 1 mM PMSF, 20 mM sodium fluoride, 50 nM Microcystin, one Protease inhibitor cocktail tablet per 50 ml wash buffer.
4. Shaker/rotisserie (Labquake, Krackeler Scientific, Albany, NY).

2.7. SDS-PAGE and In-Gel Tryptic Digestion

1. 4–20% NuPAGE gradient gels (Invitrogen, Carlsbad, CA).
2. Sypro Ruby Stain (Invitrogen).
3. Dark Reader transilluminator (Clare Chemical, Dolores, CO).
4. Razor blade.
5. Sequencing grade trypsin (Promega TPCK-treated, Cat#: V5111).
6. Ammonium bicarbonate, acetonitrile, DTT, and iodoacetamide.
7. Vacuum centrifuge.

2.8. Immobilized Metal Ion Affinity Chromatography

1. Resuspension/wash buffer: 30% acetonitrile, 0.25 M glacial acetic acid, and HPLC-grade water.
2. Elution buffer: 13.5 μl ammonium hydroxide and 486.5 μl HPLC water.
3. Water (18 MΩ) was purified using a Barnstead Nanopure system (Thermo Fisher Scientific).
4. PhosSelect iron affinity gel (Sigma #P9740).
5. TipOne 0.5–20 μl filter tips (#1121-4810, USA Scientific, Ocala, FL).
6. Variable speed microcentrifuge and vacuum centrifuge.
7. 1.5-ml microcentrifuge tubes.

2.9. LC/MS^E and LC/MS/MS Analysis

1. Acetonitrile (CHROMASOLV HPLC gradient grade, ≥99.9%) and formic acid (ACS regent grade).
2. Water (18 MΩ) was distilled and purified using a High-Q 103S water purification system (Wilmette, IL).
3. Mobile phases: (A) 99.9% water and 0.1% formic acid and (B) 99.9% acetonitrile and 0.1% formic acid.
4. Glu-fibrinopeptide B (Sigma-Aldrich).
5. Symmetry C18 trapping column (internal diameter 180 μm, and length 20 mm) (Waters Corporation, Milford, MA).
6. Bridged-ethyl hybrid (BEH) C18 reversed-phase column (1.7 μm particle size) with an internal diameter of 75 μm and length of 250 mm (Waters Corporation).
7. nanoACQUITY ultra-performance liquid chromatograph (UPLC) (Waters Corporation) coupled to a Q-Tof Premier mass spectrometer (Waters Corporation).
8. MassLynx (version 4.1, Waters Corporation).

2.10. Processing and Database Searching of LC/MS^E and LC/MS/MS Datasets

1. ProteinLynx Global Server 2.4 (PLGS) software with the IDENTITY^E (Ion Accounting) search algorithm. (Waters Corporation).
2. Mascot (version 2.2.03, Matrix Sciences, www.matrixscience.com).
3. Computer system capable of storing, processing, and analyzing data using PLGS and Mascot.

3. Methods

We developed a high-throughput protocol for analyzing LRR RLK phosphorylation sites using a Premier Q-ToF mass spectrometer functioning in both data-dependent acquisition LC/MS/MS and data-independent LC/MS^E modes (Fig. 1). Due to the ability to easily overexpress and purify His-tagged proteins in *E. coli*, over 100 His_7-tagged LRR RLK cytoplasmic domains were examined for in vitro autophosphorylation sites. We found that Ni-NTA-purified proteins could be digested with trypsin in solution and subjected to LC/MS/MS analysis without SDS-PAGE purification. Such in-solution digestion of LRR RLK cytoplasmic domains resulted in a greater number of phosphopeptide identifications than SDS-PAGE followed by in-gel digestion. For in vivo phosphorylation site determination, we used epitope-tagged LRR RLKs immunoprecipitated from 11-day-old *Arabidopsis* seedlings grown in shaking liquid culture as the source material for LC/MS/MS analysis.

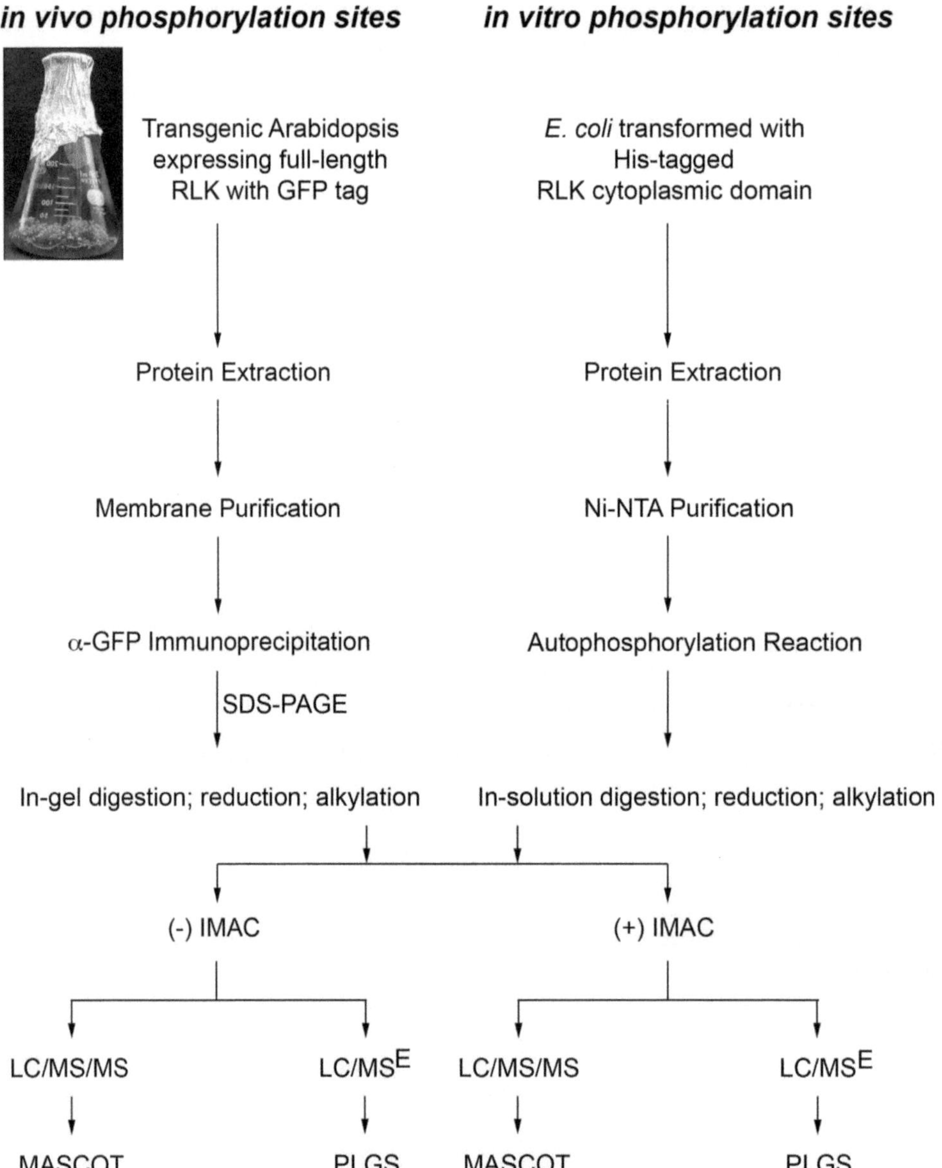

Fig. 1. Flow diagram of the protocol for isolating and processing receptor kinase protein from microsomal membrane fractions of *Arabidopsis* plants or recombinant protein from *Escherichia coli*, prior to LC/MS/MS analysis by two independent methods.

We have found in our previous work that separate runs with and without immobilized metal ion affinity chromatography (IMAC) yield the maximum number of phosphorylation sites (12, 15). All samples were additionally analyzed by two LC/MS/MS approaches: data-independent acquisition LC/MS^E and data-dependent acquisition LC/MS/MS. LC/MS^E is a novel mode of generating product ion data for all coeluting precursors in parallel as opposed to LC/MS/MS where coeluting precursors must be serially fragmented one at a time. The differences between each method as it relates to protein and proteome characterization have been experimentally described in detail (16). Due to their unique analytical features, both approaches were used to characterize LRR RLK phosphorylation sites. LC/MS^E is particularly useful for identification of low-abundance phosphopeptides.

3.1. Recombinant Protein Expression, Extraction, and Purification Using Ni-NTA Beads

1. A portion of an overnight culture is transferred to a 250-ml Erlenmayer flask containing 50 ml LB medium with appropriate antibiotics. Grow the culture at 37°C for 2 h and check the OD_{600} at regular intervals. Induce the culture with 0.5 mM IPTG and continue to grow at 28°C for 4 h (OD_{600} should be between 0.6 and 0.8).
2. Harvest the cells by centrifuging the 50 ml culture at 4,000 × *g* for 5 min. Resuspend the pellet in 2 ml lysis buffer and sonicate on ice with three 20-s pulses. Centrifuge at 21,000 × *g* for 20 min at 4°C. Collect the supernatant and store on ice.
3. Equilibrate 50 μl Ni-NTA beads with 250 μl equilibration buffer. Centrifuge the beads at 4,000 × *g* for 5 min. Discard the supernatant and add 250 μl equilibration buffer. Repeat the process three times.
4. Add the equilibrated beads to the supernatant collected from Subheading 3.1, step 2. Agitate for 1 h at 4°C and centrifuge at 4,000 × *g* for 5 min. Discard the supernatant (unbound material). Add 500 μl wash buffer and centrifuge beads at 4,000 × *g* for 5 min. Repeat the wash process three times. Elute the protein with 100 μl elution buffer by adding buffer to the washed beads and agitating the tube for 3 min. Centrifuge at 4,000 × *g* for 1 min and collect the supernatant. Repeat the bead elution process three more times, collecting each eluate in a separate tube. Measure the protein amount in each fraction using the Bradford assay kit.

3.2. In Vitro Kinase Reaction Using Unlabeled ATP

1. Autophosphorylation of purified recombinant kinase proceeds in a 20 μl reaction volume with 1× kinase buffer and 2 μg recombinant protein. The reaction is incubated at 25°C for 1 h.

3.3. In-Solution Tryptic Digestion

1. After the completion of the kinase reaction in Subheading 3.2, step 1, add 30-fold molar excess TCEP (assuming average protein mass of 30 kDa and six Cys residues per protein) and incubate at 37°C for 1 h. Proteins are subsequently alkylated using a 30-fold molar excess of iodoacetamide for 1 h in the dark at room temperature. The protein in solution is digested with trypsin overnight at 37°C using a 1:10 trypsin:protein ratio.

2. Once the digestion is complete, 1/10 volume of chloroform is added and the sample is vortexed for 30 s. Centrifuge at 21,000 × *g* for 2 min at 4°C. The upper aqueous layer containing the peptides is removed, taken to dryness in a vacuum centrifuge, and stored at −80°C.

3.4. Purification of Microsomal Membrane Fractions from Plant Tissues

1. *Arabidopsis* seedlings are obtained by sterilizing 40-mg seeds in ethanol followed by washing with 30% bleach solution for 20 min, and then extensive washing with water. Vernalize seeds at 4°C for 48 h, then transfer to a 250-ml Erlenmeyer flask containing 40 ml of sterile MS medium, and shake at 80 rpm under constant white light for 11 days.
2. Remove excess media from seedlings and homogenize approximately 60-g plant material with a blender using 120 ml of cold extraction buffer for up to 10 min until a homogenous mixture is obtained (see Note 1).
3. Centrifuge at 6,000 × *g* for 15 min at 4°C and transfer the supernatant into polypropylene centrifuge tubes (see Note 2).
4. Perform ultracentrifugation at 100,000 × *g* for 2 h at 4°C. The compact pellet obtained is the microsomal fraction.

3.5. Membrane Protein Extraction

1. Resuspend the pellet in 3 ml resuspension buffer with scraping to bring the pellet into solution. Sonicate the resuspended extract and rock the extract for 15 min on a shaker at 4°C (see Note 3).
2. Centrifuge at 21,000 × *g* for 20 min at 4°C and transfer the supernatant to a fresh tube.
3. Measure the protein amount using the Bradford assay kit.

3.6. Immunoprecipitation

1. Adjust the protein concentration from Subheading 3.5, step 3, to 1 mg/ml using dilution buffer (see Note 4).
2. Add 500 μl resuspension buffer to 100 μl FLAG M_2 beads and centrifuge at 4,000 × *g* for 1 min. Repeat the process three times.
3. Add the washed FLAG M_2 beads to the diluted protein extract from Subheading 3.6, step 1, and agitate overnight on a shaker at 4°C.
4. The next day, centrifuge the beads at 4,000 × *g* for 5 min at 4°C. Resuspend the beads in 500 μl wash buffer. Centrifuge at 4,000 × *g* for 1 min at 4°C. Repeat the process three times.
5. Elute the protein from the beads by adding 50 μl of 2× SDS-PAGE loading buffer and boil in a water bath for 5 min. Centrifuge at 4,000 × *g* for 1 min and collect the supernatant in a separate microfuge tube (eluate 1). Repeat the process three more times and collect the eluates in separate microfuge tubes.
6. Store the eluted fractions at −80°C or proceed directly to SDS-PAGE.

3.7. SDS-PAGE and In-Gel Tryptic Digestion

1. Separate the samples from Subheading 3.6, step 6, on an SDS-PAGE gel using standard protocols. NuPAGE precast gels from Invitrogen are compatible with downstream LC/MS/MS procedures and have worked well for this step in our hands. Follow the manufacturer's instructions (see Note 5).
2. After the run, fix the gel in 10% methanol:7% acetic acid for 1 h in a plastic box (see Note 6).
3. Stain the gel with Sypro Ruby in the same plastic box overnight on a shaker.
4. Red bands are visualized on the gel using the Dark Reader blue light transilluminator.
5. Cut the gel slice and store in deionized water at 4°C (see Note 7). Wash gel pieces using 500 μl of 50 mM ammonium bicarbonate in 50% acetonitrile for 15 min, with gentle agitation (vortex at the lowest setting). Discard wash solution. Repeat wash at least two times. Most of the stain should have been removed from the gel pieces (see Note 8).
6. Rinse gel pieces briefly with 500 μl of 100% acetonitrile and discard the rinse solution. Dehydrate the gel pieces with 500 μl of 100% acetonitrile for 20 min at room temperature with gentle agitation. Discard acetonitrile, and allow gel pieces to air dry.
7. Reduce the in-gel protein with 150 μl 10 mM DTT in 100 mM ammonium bicarbonate for 30 min at 56°C. Cool the sample to room temperature, and remove and discard DTT solution.
8. Alkylate the in-gel protein with 100 μl 50 mM iodoacetamide in 100 mM ammonium bicarbonate *in the dark* at room temperature for 30 min. Remove solution (see Note 9).
9. Wash the gel pieces at room temperature for 15 min with 500 μl of 50 mM ammonium bicarbonate in 50% acetonitrile. Rinse gel pieces briefly with 500 μl 100% acetonitrile and discard solution. Dehydrate the gel pieces for 20 min at room temperature with 500 μl of 100% acetonitrile. Discard acetonitrile.
10. Add 30 μl of 0.02 μg/μl trypsin in 40 mM ammonium bicarbonate (see Note 10). Wrap the tube in aluminum foil and place the tube in a rack in a water bath at 37°C for 16–18 h (see Note 11).
11. After digestion, spin in a microcentrifuge for 15 s to deposit all liquid in the bottom of the tube and transfer supernatant to a fresh tube on ice. Add 25–50 μl of extraction solution (60% acetonitrile and 1% TFA) to the remaining gel pieces followed by vortexing at the lowest setting (see Note 12). Spin in a microcentrifuge for 15 s and add the supernatant (containing additional tryptic peptides) to the original digestion solution tube on ice. Extract the gel pieces again with an additional

25–50 μl of extraction solution. Spin down sample and transfer the supernatant to the original digestion solution tube on ice.

12. Evaporate the pooled peptides to dryness in a vacuum centrifuge for immediate LC/MS/MS or freeze in liquid nitrogen and store at −80°C for future use.

3.8. Immobilized Metal Ion Affinity Chromatography

1. Prepare 0.5–20-μl filter tips by cutting the end off with a razor blade. Bore a hole in the lid of a 1.5-ml microfuge tube with a cork borer. Close the lid and insert cut filter tip into the hole. Add 50 μl of wash buffer and spin at 1,500 × *g* for 30 s.
2. Remove PhosSelect resin from −20°C storage and mix gently. Place 10 μl of resin in a fresh microfuge tube and add 100 μl wash buffer. Spin at 1,500 × *g* for 2 min at room temp. Repeat the wash two more times and keep beads on ice.
3. Resuspend dried peptides from Subheading 3.3, step 2, or 3.7, step 12, with 40 μl wash buffer. Add 10 μl of washed PhosSelect beads and incubate the mixture at room temp for 1 h on a vortex mixer at minimum speed.
4. After the binding reaction is complete, centrifuge the beads at 1,500 × *g* for 30 s. Remove supernatant (save if you want to do LC/MS/MS on unbound fraction). Pipet the resin mix on the top of the prepared filter tip from Subheading 3.8, step 1. Centrifuge at 1,500 × *g* for 30 s. Wash the beads with 10 μl of wash buffer. Spin the tip at 1,500 × *g* for 30 s. Wash the beads with 50 μl of wash buffer. Centrifuge the tip at 1,500 × *g* for 30 s. Wash the beads a final time with 50 μl of HPLC water and centrifuge the tip at 1,500 × *g* for 30 s.
5. Elute the peptides from the beads with 50 μl of elution buffer (see Note 13). Repeat the elution two times. Pool all three 50 μl eluted fractions into one tube and dry in a vacuum centrifuge. Proceed directly to LC/MS/MS or store peptides at −80°C for future use.

3.9. LC/MS/MS and LC/MS^E Analysis

1. To prepare the sample for analysis, solubilize the dried peptides from Subheadings 3.3, step 2, 3.7, step 12, or 3.8, step 5, in 100% Mobile Phase A to produce a concentration suitable for loading onto the reversed-phase column. To facilitate solubilization of the peptides, the sample can be vortexed with intermittent sonication using a sonicating water bath (see Note 14).
2. Using a binary solvent system comprising 99.9% water and 0.1% formic acid (Mobile Phase A) and 99.9% acetonitrile and 0.1% formic acid (Mobile Phase B), equilibrate the trap and column with 2% Mobile Phase B. Inject each sample (typically 5 μl) and preconcentrate the peptides online at a flow rate of 10 μl/min for 4 min using the Symmetry C18 trapping column. Once loading and desalting are complete, adjust the

flow rate to 300 nl/min and then switch the flow to the BEH C18 reversed-phase column.

3. For peptide separation and elution into the NanoLockSpray ion source, use a linear gradient of 2–40% of Mobile Phase B over 30 min (see Note 15).
4. To perform data analysis with high mass measurement accuracy, use 100 fmol/μl of glu-fibrinopeptide B as the lockmass calibrant. Introduce this peptide into the NanoLockSpray ion source at a flow rate of 600 nl/min and enable this calibrant to be sampled during the acquisition every 30 s (see Note 16).
5. Use the V-mode to enable a mass resolving power of 10,000 full width at half height (FWHH) (see Note 17).
6. For LC/MS^E analysis, data are collected utilizing two scanning methods of MS analysis, each over an m/z range of 50–1,990 using the "expression" mode that acquires alternating 2-s scans of normal and elevated collision energy (17, 18). For this acquisition, the data are collected at a constant collision energy setting of 4 V during low-energy MS mode scans, whereas a step from 15 to 30 V of collision energy is used during the high-energy MS^E mode scans. In this manner, all peptides are selected for fragmentation.
7. For LC/MS/MS analysis, the same amount of sample is used as in the LC/MS^E analysis; however, the data generated by the mass spectrometer is based on intensity-driven parameters unique to each instrument. For the nanoACQUITY-Q-Tof Premier, typical switching parameters were used (see Note 18). Using dynamic exclusion to minimize multiple MS/MS events for the same precursor ion, set the acquisition to perform MS scans (m/z 400–1,990) of 1.3 s for peptide detection with 2-s MS/MS scans for detected precursors. To obtain the best balance of duty cycle and product ion spectral quality, select a maximum of eight precursor ions to be selected per MS/MS switching event, with up to two MS/MS scans allowed per precursor ion interrogated. Collision energies used for precursor fragmentation are determined by the instrument according to the selected precursor m/z and its charge state.

3.10. Processing and Database Searching of LC/MS^E and LC/MS/MS Datasets

1. Process LC/MS^E and LC/MS/MS raw data files with PLGS 2.4 to generate product ion spectra for subsequent database searching using the Ion Accounting algorithm within PLGS (see Note 19) and output pkl files for subsequent database searching using Mascot, respectively.
2. For database searching, use the most current *Arabidopsis* protein database available at The Arabidopsis Information Resource (TAIR) Web site (www.arabidopsis.org) with a fixed carbamidomethyl modification for Cys residues and variable

modifications for Met oxidation, Asn and Gln deamidation, N-terminal acetylation, and phosphorylation of Ser, Thr, and Tyr residues.

3. For ion accounting analysis of LC/MS^E data, use the following search parameters (settings in parentheses): precursor and product ion tolerance (automatic setting), minimum number of peptide matches (1), minimum number of product ion matches per peptide (3), minimum number of product ion matches per protein (7), maximum number of missed tryptic cleavage sites (1), and maximum false-positive rate (FPR) (2%) (see Note 20). All phosphopeptide matches obtained under the 2% maximum FPR require manual inspection to determine correct assignments; all other peptides below the 2% FPR can be accepted as correct matches.
4. For Mascot analysis of LC/MS/MS data, use mass tolerances of 50 ppm and 0.05 Da for precursor and product ions, respectively. All data are searched against the randomized protein database, and the FPR for identification is calculated based on the number of peptide matches in the forward versus the randomized database. All phosphopeptide matches with Mascot scores of at least 25 require manual inspection to determine correct assignments; all other peptide matches with Mascot scores exceeding the 95% confidence level score can be accepted as correct matches.

4. Notes

1. Blend for short pulses of 30 s followed by a 30-s pause to prevent the sample from overheating.
2. While transferring the supernatant into polypropylene tubes, pass it through Miracloth (Calbiochem) to remove debris.
3. The samples should be on ice while performing sonication to prevent overheating.
4. The protein final concentration should be 1 mg/ml with a detergent concentration that is less than 0.2%.
5. Gel staining and preparation of peptides must be performed with labware that has never been in contact with nonfat milk, BSA, or any other protein-blocking agent to prevent carryover contamination.
6. Sypro Ruby is quenched by glass. Always use plastic trays for staining purposes.
7. Always use powder-free gloves when handling samples. Keratin and latex proteins are potential sources of contamination.

8. Prepare ammonium bicarbonate buffer and all reagent solutions on the day that they are to be used. Prepare DTT, iodoacetamide, and diluted trypsin solutions just before addition to the samples.
9. Be sure to wear gloves while handling iodoacetamide. When finished with iodoacetamide solutions, neutralize with a two-fold molar excess of DTT and discard.
10. Be sure that enough volume is added to ensure complete rehydration of gel pieces. More than 30 μl trypsin solution may be needed to completely rehydrate pieces of gel from a large band. Promega trypsin is sold as 20 μg dried protein/vial. Reconstitute the trypsin at 1 μg/μl in the 50 mM acetic acid solution shipped along with the enzyme. Freeze this solution at −80°C in aliquots. Thaw and dilute the required amount of stock solution in the digestion buffer just before needed. Note that Promega *Mass Spec grade* trypsin may be used.
11. The foil wrap helps minimize the amount of condensate that collects inside the reaction tube cap during the incubation and, thus, prevent the gel pieces from drying out overnight.
12. Extraction solution is made by combining 600 μl 100% ACN, 300 μl fresh HPLC-grade H_2O, and 100 μl of a fresh 10% TFA aqueous stock solution. Use the smallest volume of extraction solution possible to minimize dilution of the peptides. Extract for at least 10 min.
13. Add 50 μl of elution buffer to the PhosSelect beads and agitate it on a thermomixer (Eppendorf) at 300 setting for 5 min. Spin down at 1,500 × *g* and collect the supernatant. Repeat the entire process two times.
14. If additional sonication is not helpful, the solution can be adjusted to contain a certain percentage of the organic Mobile Phase B. However, increasing the organic composition to greater than 10% can cause hydrophilic peptides to flow through the trapping column. In most cases, all LRR RLK samples were readily soluble in Mobile Phase A. If any components remain insoluble or not, the sample should be filtered using a pipette tip containing a porous filter to remove any particulates in order to avoid clogging of the nanoLC system.
15. The 30-min gradient is sufficient to separate peptides of less complex mixtures, such as the immunoprecipitated LRR RLKs. To promote potentially more peptide identifications by LC/MS/MS, a 60-min gradient can be used; however, this will limit the number of samples analyzed for a given amount of instrument time.

16. With the use of lock mass, the mass measurement accuracies for precursor ion and product ions are typically less than 10 and 20 ppm, respectively.
17. Although the W-mode of analysis can be used to achieve a resolving power of 20,000 FWHH, this increases scanning rates, thus lowering the number of collision-induced dissociation (CID) events that can be performed compared to using the V-mode. Based on the charge state of peptides obtained from tryptically digested proteins, a resolving power of 10,000 FWHH is sufficient for charge state determination of an $[M+4H]^{4+}$ ion in the m/z range of 400–2,000.
18. Typical switching parameters for a mass spectrometer can be defined for a data-dependent LC/MS/MS acquisition as those which balance the duty cycle between the number of precursor ions interrogated and the product ion spectral quality.
19. The ion accounting search algorithm was specifically developed for searching data-independent MS^E datasets ($IDENTITY^E$) as described by Li et al. (19).
20. The FPR is calculated during ion accounting search depletion loops based on the appearance of random matches observed when searching a concatenated forward and its corresponding randomized database (19).

Acknowledgments

This work was supported by National Science Foundation grants MCB-1021363 and MCB-0740211. We also thank the research agencies of North Carolina State University and the North Carolina Agricultural Research Service for continued support of our biological mass spectrometry research.

References

1. Hubbard SR, Miller WT (2007) Receptor tyrosine kinases: mechanisms of activation and signaling. Curr Opin Cell Biol 19:117–123
2. Huse M, Kuriyan J (2002) The conformational plasticity of protein kinases. Cell 109:275–282
3. Kim TW, Wang ZY (2010) Brassinosteroid signal transduction from receptor kinases to transcription factors. Annu Rev Plant Biol 61:681–704
4. Vert G, Nemhauser JL, Geldner N, Hong F, Chory J (2005) Molecular mechanisms of steroid hormone signaling in plants. Annu Rev Cell Dev Biol 21:177–201
5. Clouse SD, Goshe MB, Huber SC, Li J (2008) Functional analysis and phosphorylation site mapping of leucine-rich repeat receptor-like kinases. In: Agrawal GK, Rakwal R (eds) Plant proteomics: technologies, strategies and applications. John Wiley & Sons, New York, pp 469–484
6. Gou X, He K, Yang H, Yuan T, Lin H, Clouse SD, Li J (2010) Genome-wide cloning and sequence analysis of leucine-rich repeat receptor-like protein kinase genes in *Arabidopsis thaliana*. BMC Genomics 11:19

7. Li J, Chory J (1997) A putative leucine-rich repeat receptor kinase involved in brassinsteroid signal transduction. Cell 90:929–938
8. Li J, Wen J, Lease KA, Doke JT, Tax FE, Walker JC (2002) BAK1, an Arabidopsis LRR receptor-like protein kinase, interacts with BRI1 and modulates brassinosteroid signaling. Cell 110:213–222
9. Nam KH, Li J (2002) BRI1/BAK1, a receptor kinase pair mediating brassinosteroid signaling. Cell 110:203–212
10. Oh MH, Ray WK, Huber SC, Asara JM, Gage DA, Clouse SD (2000) Recombinant brassinosteroid insensitive 1 receptor-like kinase autophosphorylates on serine and threonine residues and phosphorylates a conserved peptide motif *in vitro*. Plant Physiol 124:751–766
11. Wang ZY, Seto H, Fujioka S, Yoshida S, Chory J (2001) BRI1 is a critical component of a plasma-membrane receptor for plant steroids. Nature 410:380–383
12. Wang X, Kota U, He K, Blackburn K, Li J, Goshe MB, Huber SC, Clouse SD (2008) Sequential transphosphorylation of the BRI1/BAK1 receptor kinase complex impacts early events in brassinosteroid signaling. Dev Cell 15:220–235
13. Wang X, Li X, Meisenhelder J, Hunter T, Yoshida S, Asami T, Chory J (2005) Autoregulation and homodimerization are involved in the activation of the plant steroid receptor BRI1. Dev Cell 8:855–865
14. Oh MH, Wang X, Kota U, Goshe MB, Clouse SD, Huber SC (2009) Tyrosine phosphorylation of the BRI1 receptor kinase emerges as a component of brassinosteroid signaling in Arabidopsis. Proc Natl Acad Sci USA 106:658–663
15. Wang X, Goshe MB, Soderblom EJ, Phinney BS, Kuchar JA, Li J, Asami T, Yoshida S, Huber SC, Clouse SD (2005) Identification and functional analysis of *in vivo* phosphorylation sites of the Arabidopsis BRASSINOSTEROID-INSENSITIVE1 receptor kinase. Plant Cell 17:1685–1703
16. Blackburn K, Mbeunkui F, Mitra SK, Mentzel T, Goshe MB (2010) Improving protein and proteome coverage through data-independent multiplexed peptide fragmentation. J Proteome Res 9:3621–3637
17. Silva JC, Denny R, Dorschel C, Gorenstein MV, Li GZ, Richardson K, Wall D, Geromanos SJ (2006) Simultaneous qualitative and quantitative analysis of the *Escherichia coli* proteome: a sweet tale. Mol Cell Proteomics 5:589–607
18. Silva JC, Denny R, Dorschel CA, Gorenstein M, Kass IJ, Li GZ, McKenna T, Nold MJ, Richardson K, Young P, Geromanos S (2005) Quantitative proteomic analysis by accurate mass retention time pairs. Anal Chem 77:2187–2200
19. Li GZ, Vissers JP, Silva JC, Golick D, Gorenstein MV, Geromanos SJ (2009) Database searching and accounting of multiplexed precursor and product ion spectra from the data independent analysis of simple and complex peptide mixtures. Proteomics 9:1696–1719

Chapter 2

Quantitative Measurement of Phosphopeptides and Proteins via Stable Isotope Labeling in *Arabidopsis* and Functional Phosphoproteomic Strategies

Ning Li

Abstract

Protein phosphorylation is one type of posttranslational modification, which regulates a large number of cellular processes in plant cells. As an emerging powerful biotechnology that integrates all aspects of advantages from mass spectrometry, bioinformatics, and genomics, phosphoproteomics offers us an unprecedented high-throughput methodology with high sensitivity and dashing speed in identifying a large complement of phosphoproteins from plant cells within a relatively short period of time. Needless to say, phosphoproteomics has become an integral portion of life sciences, which penetrates various research disciplines of biology, agriculture, and forestry and irreversibly changes the way by which plant scientists study biological problems.

Because phosphorylation/dephosphorylation of protein is dynamic in cells and the amount of phosphoproteins is low, the preservation of a phosphor group onto phosphosite throughout protein purification as well as enrichment of these phosphoproteins during purification has become a serious technical issue. To overcome difficulties commonly associated with phosphoprotein isolation, phosphopeptides' enrichment, and mass spectrometry analysis, we have developed a urea-based phosphoprotein purification protocol for plants, which instantly denatures plant proteins once the total cell content comes into contact with the UEB solution. To measure the alteration of phosphorylation on a phosphosite using mass spectrometer, an in vivo ^{15}N metabolic labeling method (SILIA, i.e., stable isotope labeling in *Arabidopsis*) has been developed and applied for *Arabidopsis* differential phosphoproteomics. Thus far, hundreds of signaling-specific phosphoproteins have been identified using both *label-free* and ^{15}N-labeled differential phosphoproteomic approach. The phosphoproteomics has allowed us to identify a number of signaling components mediating plant cell signaling in *Arabidopsis*. It is envisaged that a huge number of phosphosites will continue to be uncovered from phosphoproteomics in the near future, which will become instrumental for the development of plant phosphor-relay networks and molecular systems biology.

Key words: Plant, Functional phosphoproteomics, Mass spectrometry, Stable isotope labeling in *Arabidopsis*, In vivo stable isotope ^{15}N labeling, Quantitative proteomics, Site-directed mutagenesis

Zhi-Yong Wang and Zhenbiao Yang (eds.), *Plant Signalling Networks: Methods and Protocols*, Methods in Molecular Biology, vol. 876, DOI 10.1007/978-1-61779-809-2_2,

1. Introduction

Reversible protein phosphorylation plays a central role in cell signaling, regulation of gene expression, controlling of the growth and development of an organism, and its adaptation to environmental changes (1). Plants also make use of phosphor-relay mechanism for ethylene signaling (2). Phosphoproteomics (3) has been developed for profiling global protein phosphorylation at a given developmental stage or in response to a specific external cue. Identification of protein phosphorylaiton has always been technically difficult in the past due to the relatively low abundance as well as the labile nature of the phosphorylation site. With the emerging powerful phosphoproteomic technology, i.e., the immobilized metal-ion affinity chromatography (IMAC)-based phosphopeptides' enrichment coupled with liquid chromatography mass spectrometric sequencing (LC–MS/MS) of phosphorylated peptides, we are now able to profile phosphoproteins at large scale and determine the phosphorylation sites associated with a developmental cue or an environmental inducer (3–8). Ever since Nühse et al. (4) have identified more than 300 phosphorylation sites from *Arabidopsis* membrane proteins using the phosphoproteomic approach, nearly 30,430 phosphopeptides have been characterized thus far from the model plant *Arabidopsis* according to PhosphAT3.0 (http://phosphat.mpimp-golm.mpg.de/statistics.html).

With the advent of breakthroughs in quantitative differential proteomics, i.e., the *label-free* approach (9–11) and the isotope-assisted approach (12–18), the mass spectrometry-based quantitation of phosphorylation level of a large number of phosphosites has become the focus of quantitative phosphoproteomics. The advantages of the in vitro isotope-labeling methods, such as isotope-coded affinity tags (ICATs) (12), isotope tagging for relative and absolute protein quantitation (iTRAQ) (17), and ^{18}O-enriched water (13, 19), are well recognized because they are quite versatile and readily used to incorporate peptides isolated from virtually any proteins despite their known shortcomings, such as their susceptibility to sample manipulation error (20–23).

To further advance the study on quantitative proteomics and phosphoproteomics and to efficiently measure the phosphorylation levels of a large number of phosphosites using mass spectrometer, an in vivo stable isotope-labeling method has been introduced into this field. The differential peptide abundance can be measured through in vivo metabolic ^{15}N labeling, in which a heavy isotopic tag, either ^{15}N or ^{13}C, in the form of salt, amino acid, or sugar, is mixed into the food or medium for an organism. When this organism such as *Arabidopsis* grows on the labeled supporting media, it presumably assimilates the heavy stable isotopes into the entire protein complement. Because the in vivo metabolic labeling has little measureable detrimental effect on the growth and development of an organism, this in vivo labeling approach can be especially useful for experiments involving smaller model plants, such as *Arabidopsis*. The key advantage of this in vivo labeling approach is that the mixing of a pair of in vivo metabolically labeled plant protein samples at the earliest step of manipulation possible eliminates the deviation of a peptide ion measurement resulting from multiple steps of sample manipulation throughout an extensive peptide preparation process. Stable isotope labeling with amino acids in cell culture (SILAC) (14) is an in vivo metabolic labeling technique used frequently, in which isotope-coded amino acids (such as Lys or Arg) labeled with ^{15}N or ^{13}C are incorporated into proteins of an organism (24–26). Alternatively, ^{15}N- or ^{13}C-labeled salts or sugars are being incorporated into the organism studied (27). This type of stable isotope metabolic-labeling approach has been applied successfully onto numerous model organisms for quantitative proteomic studies (9, 18, 28–34). Moreover, ^{15}N metabolic labeling has been applied for top-down proteomics and serves as a standard to evaluate other quantitative proteomics techniques, such as DIGE (35) and spectral counting (36). In some cases, both ^{13}C and ^{15}N labeling are combined together to measure protein abundance of three different biological samples (37).

Thus, the advancement in quantitative proteomics in general and phosphoproteomics in specific prompts us to establish a practical protocol for the study of differential phosphoproteomics. Our protocol integrates the recent advancement in proteomics and depicts the processes of both *label-free* (11) and stable isotope labeling in *Arabidopsis* (SILIA) labeling methods. Both are specially designed for *Arabidopsis* plant growing on the solid agar medium because this solid medium is a common growth condition widely used by many laboratories around the world to carry out *Arabidopsis* mutant screens and physiological studies. This protocol should be a useful application of functional phosphoproteomics in plant cell signaling and pioneer an alternative workflow, in addition to genetic screening of *Arabidopsis* mutants,for identification of signaling components mediating the intricate phosphor-relay network during plant cell signaling: i.e., *MS/MS- and bioinformatics-based phosphosite identification → validation by in vitro kinase assay → in planta validation of the functional role of putative phosphorylation site by site-directed mutagenesis.*

2. Materials

2.1. Plant Growth and Harvest

1. MS medium for *label-free* plant growth (11): Murashige and Skoog (MS) basal salt mixture (Sigma-Aldrich, St. Louis, MO) 4.33 g/L, sucrose 10 g/L, 1 mg/L thiamine HCL, 0.1 mg/L

pyridoxine, 0.1 mg/L nicotinic acid, 100 mg/L myo-inositol, and 0.8% bacteriological agar. Adjust pH to 5.7 by KOH.

2. Plant growth medium for in vivo SILIA (38): 9 mM KNO_3 or $K^{15}NO_3$, 0.4 mM $Ca_5OH(PO_4)_3$, 2 mM $MgSO_4$, 1.3 mM H_3PO_4, 50 μM Fe-EDTA, 70 μM H_3BO_3, 14 μM $MnCl_2$, 0.5 μM $CuSO_4$, 1 μM $ZnSO_4$, 0.2 μM Na_2MoO_4, 10 μM NaCl, 0.01 μM $CoCl_2$, 10 g/L sucrose, 1 mg/L thiamine HCL, 0.1 mg/L pyridoxine, 0.1 mg/L nicotinic acid, 100 mg/L myo-inositol, and 0.8% bacteriological agar. Adjust pH to 5.7 by KOH.

This medium formula was modified from MS medium and specially designed for *Arabidopsis* growing on solid medium in plate or jar. The medium should be made in two separate sets and labeled clearly as a light nitrogen medium (^{14}N, from normal KNO_3) or a heavy nitrogen medium (^{15}N, from $K^{15}NO_3$).

2.2. Protein Sample Preparations

1. Urea extraction buffer (UEB) is designed for the initial dissolving of plant cell lysate (38), which is 150 mM Tris–HCl, pH 7.6, 8 M urea, 0.5% SDS, 1.2% TritonX-100, 20 mM EDTA, 20 mM EGTA, 50 mM NaF, 2 mM $NaVO_3$, and 1% Glycerol 2-phosphate disodium salt hydrate, stored at 4°C. Add the following compounds immediately to make a final concentration of 1 mM PMSF, 5 mM DTT, 0.5% *phosphatase inhibitors cocktail 2* (and *phosphatase inhibitors cocktail 1*, which can be purchased only in some regions and countries), 1× complete EDTA-free protease inhibitors cocktail, 5 mM ascorbic acid, and 2% PVPP. The final UEB solution looks brownish and can be stored at −80°C for experimental use for at least 3 months.
2. Resuspension buffer (RSB): 50 mM Tris–HCl, pH 7.6, 8 M urea, 10 mM DTT, 1% SDS, and 10 mM EDTA. Stored at 4°C.
3. Precipitation solution: Acetone:methanol (12:1), precooled at −20°C.
4. Rinse solution: Acetone:methanol:H_2O (12:1:1.4), precooled at −20°C.

2.3. SDS-PAGE and In-Gel Digestion

1. Coomassie staining buffer: 0.2% Brilliant Blue G250 in 20% methanol, 0.5% acetic acid.
2. Destain buffer: 20% methanol, 0.5% acetic acid.
3. Wash buffer: 50% acetonitrile (ACN)/25 mM NH_4HCO_3.
4. DTT solution: 10 mM DTT in Milli-Q water.
5. IAA solution: 55 mM iodoacetamide (IAA) in 25 mM NH_4HCO_3. Use freshly made solution every time and keep solution in dark.

6. Trypsin solution: 30 ng/μl TPCK-treated trypsin in 25 mM NH_4HCO_3.
7. Extraction buffer: 1% formic acid in 50% ACN.

2.4. Ion Exchange Chromatography and IMAC/TiO_2 Enrichment

1. Ion exchange chromatography (SCX) buffer set: (A) 5 mM KH_2PO_4, pH 2.65, 30% ACN (*v*/*v*), (B) 5 mM KH_2PO_4, 350 mM KCl, pH 2.65, and 30% ACN (*v*/*v*).
2. Ion exchange gradient: 0–1 min, 0% Buffer B; 1–12 min, 15% Buffer B; 12–18 min, 35% Buffer B; 18–22 min, 100% Buffer B; 22–26 min, 100% Buffer B; 26–27 min, 0% Buffer B; and 27–40 min, 0% Buffer B. Flow rate: 1 ml/min.
3. C18 reverse phase (RP) column (Oasis HLB, waters) for peptide desalt.
4. NTA-agarose beads (Sigma).
5. $FeCl_3$ solution: 0.1 M $FeCl_3$ in Milli-Q water, freshly made before use.
6. IMAC loading buffer: 6% acetic acid/30% ACN. pH must be less than 3.0.
7. IMAC elution buffer: 200 mM ammonium phosphate, pH 4.5.
8. TiO_2 equilibration/washing buffer: 1 M glycolic acid, 5% TFA, 80% ACN.
9. TiO_2 elution buffer: 1% ammonium hydroxide.
10. TiO_2 beads (GL science Inc, Tokyo, Japan).

2.5. Ziptip and LC–MS/MS Data Acquisition/Analysis

1. Ziptip equilibrium/wash solution: 0.1% formic acid.
2. Ziptip elution solution: 1% formic acid in 50% methanol.
3. Buffer A for LC/MS RP column: 0.1% formic acid in water.
4. Buffer B for LC/MS RP column: 0.1% formic acid in ACN.

2.6. Buffer for In Vitro Kinase Assay

1. 20 mM HEPES, pH 7.5, 150 mM NaCl, 1% Triton X-100, 2.5 mM $Na_4P_2O_7$, 1 mM glycerophosphate, and 1 mM NaF; store in 4°C. Add 1 mM Na_2MoO_4, 1 mM Na_3VO_4, 1× complete EDTA-free protease inhibitors cocktail, and 1 mM PMSF freshly before use.
2. Activation mix: 10 ml 50% glycerol, 0.5 ml 50 mM ATP, 0.6 ml 1 M $MgCl_2$, and 0.15 ml 10 mg/ml BSA.
3. Trypsin digestion buffer: 50 mM Tris–HCl, pH 7.5, 150 mM NaCl, 0.1 mM $CaCl_2$, and 10 mM $MgCl_2$.
4. Synthetic peptides with the target sequences were synthesized by peptide synthesis companies on the market.

3. Methods

3.1. Plant Growth and Harvest

1. The *Arabidopsis* seeds are surface sterilized before imbibed at 4°C for 3 days in dark. Glass jars with 9 cm in diameter and 15 cm in height are autoclaved. Plant nutrient agar medium (40 ml) are poured into each jar and cooled to dry overnight in hood. For *label-free* method, only MS medium is used and not necessary to label the jar. For metabolic labeling with ^{15}N, jars were labeled to distinguish the ^{15}N medium from the ^{14}N medium.
2. Seeds are then suspended in 0.1% (*w*/*v*) agar and sown in rows on plant nutrient agar medium within the jar. Plant about 25 seeds in each jar. The distance between each seed is about 0.8 cm (see Note 1).
3. Jars with planted seeds are transferred to plant growth chambers (16-h light/8-h dark cycle, with constant temperature of 20°C). After 3 weeks, the seedlings are placed in airflow chamber for 5 h to eliminate endogenous ethylene (see Note 2).
4. Adjust the gas flow rate to fill one cultivation jar within 4 s. For *label-free* experiment, divide the jars into two fractions, treated and untreated, respectively, for 15 min or any other time frames. For SILIA experiment, seedlings grown on ^{14}N are treated with air and ^{15}N-labeled seedlings are treated with a hormone or an external inducer for a period of time. To avoid variance induced by different isotopic incorporation, two sets of reciprocal labeling are required.
5. Harvest the seedlings with liquid nitrogen and preserve the tissue in −80°C freezer.

3.2. Protein Sample Preparation

1. Seedlings (10 g) are ground into powders with liquid nitrogen in a precooled mortar. To effectively denature plant proteins for the purpose of freezing phosphor group onto phosphosite and to prevent in vitro nonspecific enzyme catalysis by kinase and phosphatase during protein preparation, a phosphoprotein extraction buffer UEB is employed during plant protein isolation. The tissue powder is then mixed with 50 ml UEB buffer and ground for 2 min. The cell lysate is transferred to centrifuge tubes and centrifuged at 10°C for 2 h (RCF = 110,000 × *g*, rpm = 39,359 for TST 60.4 Sorvall rotor). The high-speed centrifugation is designed to remove cell debris, DNA, cell wall, lipid, and RNA. The presence of these macromolecules interferes protein resuspension and separation on SDS-PAGE.
2. The supernatant is mixed with 3× volumes of precipitation solution and kept at −20°C for 2 h to precipitate proteins.

3. Centrifuge in Beckman JA10.5 rotor with 15,000 × g at 10°C for 20 min to pellet the protein.
4. Pour off supernatant, and rinse the pellet with 10 ml cold rinse solution to remove residue pigment and precipitated urea.
5. Pellet is dried in open air until there is no significant liquid droplet (about 10 min).
6. Protein pellet is then dissolved with 10 ml of RSB (see Note 3). Precipitate the protein again with 3× volumes of RSB at −20°C for 2 h.
7. Centrifuge with Beckman JA25.5 rotor (11,000 rpm at 10°C for 20 min) to pellet protein.
8. Pour off supernatant, and rinse the pellet with 10 ml cold rinse solution to remove residue pigment and precipitated urea. The pellet is dried in air. Resuspend the pellet in RSB with a final volume of 6 ml.
9. Freeze and keep the protein sample in −80°C. The protein concentration is determined by protein DC assay (BioRad).

3.3. SDS-PAGE and In-Gel Digestion

For *label-free* experiment, skip the sample mixing step (step 1 in this section) and start from step 2 to run the protein sample directly on SDS-PAGE gel.

1. Mix both ^{14}N- and ^{15}N-labeld protein samples extracted from ^{14}N-labeled tissue (untreated) and that from ^{15}N-labeled tissue (treated) at 1:1–1.5 ratio (depending on the actual ^{15}N labeling efficiency).
2. Load 40 mg of proteins onto four preparative SDS-PAGE (180 × 190 × 1–1.5 mm, 10%) gels evenly. For *label-free* experiment, both treated and untreated samples need to be loaded onto two separate sets of gels. Electrophoresis is stopped when the bromophenol blue dye migrates approximately 10 cm into the resolving gel.
3. Each gel is lightly stained with Coomassie blue (immersed in staining buffer for 10 min with gentle shake), destained, and fixed with destain buffer for 30 min (see Note 4).
4. Each gel is evenly cut into 5–50 strips. Identical strips from different gels are combined and further diced into 1-mm^3 cubes (see Note 5). The cubes from the same strip are collected into a 50-ml falcon tube.
5. Gel cubes are washed with 8 ml wash buffer for 15 min with shake. The wash buffer is then pipetted out and the wash step is repeated for two times.
6. Dehydrate the gel with 5 ml 100% ACN, pour off the solution, and dry the gel completely by flushing with compressed air.
7. Gel cubes are immersed in 5 ml of 10 mM DTT at 56°C for 1 h for reduction.

8. Remove DTT solution. Gel cubes are alkylated with 5 ml IAA solution at room temperature for 1 h. This step should be conducted in dark.
9. Wash the gel with 8 ml of wash buffer for additional two times. Dehydrate and dry the gel as in step 6.
10. The gel cubes is rehydrated in 2.5 ml trypsin solution on ice for 30 min (see Note 6) and then digested overnight at 37°C.
11. Sonicate the falcon for about 15 min, and then collect the supernatant. To further extract the digested peptides, 2 ml extraction buffer is added into the falcon. Falcon tube is sonicated for 15 min again and supernatant is collected. Repeat the extraction step for additional three times.
12. Pour off the supernatant of the same stripes together. The peptide solution is then flushed by compressed air until all ACN is removed from the solution.
13. The peptide solution is frozen by liquid nitrogen and concentrated by lyophilizer. The peptide powder could be stored at −80°C.

3.4. Ion Exchange Chromatography and IMAC/TiO_2 Enrichment

1. Peptide powder is reconstituted with 1.04 ml of 5 mM KH_2PO_4 (pH 2.65) and centrifuged at 4°C for 5 min with a maximum speed of a benchtop centrifuge to remove the insoluble.
2. The mixture of 20 μl is reserved for quality control and peptide concentration measurement.
3. Not more than 1.6 mg of peptide is loaded onto the ion exchange chromatography column to avoid overloading. Run the SCX in the program described in Subheading 2.4. Twelve fractions were collected from 2 to 26 min (2 min per fraction).
4. Fraction no. 1–3 and 10–12 are combined due to their low abundance of peptide. All the fractions are evaporated to remove the existence of ACN, then frozen by liquid nitrogen, and lyophilized to half of the original volume.
5. Activate the C18 RP column (Oasis) with 1 ml ACN.
6. The column is then equilibrated with 1 ml 0.1% formic acid.
7. Load the peptide samples from step 4 onto C18 column with syringe and reload for at least three times for maximum binding.
8. Wash the column with 1 ml 0.1% formic acid.
9. The peptide bound to the column is eluted with 1 ml 80% ACN/0.1% FA. Use spin vacuum to concentrate the eluted sample.
10. Pipette 200 μl NTA agarose beads into a 1.5-ml Eppendorf tube, and centrifuge at 5,000 rpm in a benchtop centrifuge for 30 s to remove storage buffer.

11. Equilibrate the beads with 1 ml 6% acetic acid. Wash the beads for 30 s and centrifuge at 5,000 rpm for 30 s. Remove the supernatant.
12. Add 1 ml 0.1 M $FeCl_3$ to the Eppendorf. Incubate the beads at 4°C with end-over-end rotation for 2 h.
13. Spin down the beads, and remove the charge buffer. Wash the beads with 1 ml 6% acetic acid for three times to remove free iron ions.
14. Equilibrate the beads with 1 ml IMAC loading buffer for three times.
15. The concentrated peptide sample from step 9 is reconstituted with Fe-IMAC loading buffer. For each milligram of peptide sample, 200 μl Fe-IMAC loading buffer is used to dissolve the sample in a new Eppendorf. The peptide is then incubated with 30 μl Fe^{3+}-NTA beads (from step 14) via a 45-min end-over-end incubation.
16. Spin down the beads. Remove and save the flow-through fraction. Fe-IMAC loading of 350 μl buffer is added to the Eppendorf to wash away the nonspecific binding peptides. Wash twice with Fe-IMAC loading buffer and once with water.
17. The phosphopeptides enriched by Fe-NTA beads are eluted with 50 μl elution buffer. Save the eluted sample in −20°C.
18. Add 800 μl of TiO_2 equilibration/washing buffer to flow-through fraction from step 16, and then incubate the solution with 5 mg of TiO_2 beads for 45 min (see Note 7).
19. The TiO_2 beads are washed twice with TiO_2 equilibration/washing buffer and once with water (200 μl for each step).
20. The enriched phosphopeptides are eluted with 50 μl elution buffer and combined with the samples obtained from step 17 and saved at −20°C before being passed through Ziptip enrichment.

3.5. Ziptip and LC–MS/MS Data Acquisition/Analysis

1. Acidify the peptide sample with 20 μl formic acid (see Note 8).
2. Aspirate 12 μl ACN into Ziptip, dispense to waste, and repeat three times.
3. Aspirate 12 μl Ziptip equilibrium/wash solution, dispense to waste, and repeat five times.
4. Aspirate the acidified sample and dispense 20 rounds for efficient binding.
5. Aspirate 12 μl Ziptip equilibrium/wash solution, dispense it to the waste, and repeat five times.
6. Elute the sample with 40 μl elution solution. Freeze and dry the sample in a spin vacuum. The sample is then reconstituted in 10 μl 0.1% formic acid and ready for LC/MS analysis.

7. LC–MS/MS is performed with a nanoflow LC (nano AcquityTM, Waters) coupled to an ESI-hybrid quadrupole time-of-flight (Q-TOF) Premier tandem mass spectrometer (Waters). The program MassLynx (version 4.1, Waters) is used for data acquisition and instrument control. A 180 μm × 20 mm Symmetry C18 trap column and 75 μm × 250 mm BEH130 C18 analytical column are used. The mobile phases are 0.1% HCOOH/H2O (A) and 0.1% HCOOH/CH3CN (B). LC gradient elution condition is initially 1–5% B (5 min), 40% B (90 min), 99% B (94–109 min), and then initial concentration 1% B (110–120 min), with a flow rate of 200 nl/min.
8. The mass spectrometer (Waters Q-TOF Premier) is operated in a positive ion mode with following basic parameters: source temperature is 80°C, capillary voltage is 2.4 kV, sample cone voltage is 35 V, and collision cell gas flow rate is 0.50 ml/min. The collision energy is variable during MS/MS scan according to z and m/z and the exact values are from factory instructions. Data-dependent analysis is set as below: 1-s MS m/z 250–2,000 and max 3-s MS/MS m/z 50–2,000 (continuum mode), 30-s dynamic exclusion. Three most abundant, +2, +3, or +4 charged ions, whose intensity rising above 40 counts/s, are selected in each MS/MS scan.
9. Raw data are processed using ProteinLynx 2.2.5 (smooth 3/2 Savitzky Golay and center 4 channels/80% centroid), and the resulting MS/MS dataset is searched against TAIR database (download from www.arabidopsis.org and specific for *Arabidopsis*) using MASCOT search engine. The settings in the workflow template are as follows: trypsin digestion with up to two missed cleavage sites are allowed; 100 ppm mass tolerance for MS precursor ions and 0.1 Da mass tolerance for MS/MS fragment ions; carbamidomethylation (C) is specified as a fixed modification and phosphorylation (S, T, and Y), deamidation (N, Q), and oxidation (M) are allowed as variable modifications. Figure 1 is an example to show the phosphopeptides discovered in a phosphoproteomics analysis (11). Figure 2 shows an example of light and heavy isoforms of a single phosphopeptide.

3.6. In Vitro Kinase Assay

The in vitro kinase assay is used as a useful tool to validate the phosphorylation sites identified by phosphoproteomics and bioinformatics-predicted sites (14). Once the phosphopeptides are identified, a short of stretch (21–30 amino acids long) of polypeptide containing the newly identified phosphosite (S, T, or Y), the highly conserved amino acid sequence motif surrounding the phosphosite, is fused to a HisTag (6× histidine) at C-terminus. This hybrid peptide is synthesized chemically and used as a substrate for in vitro plant kinase assay.

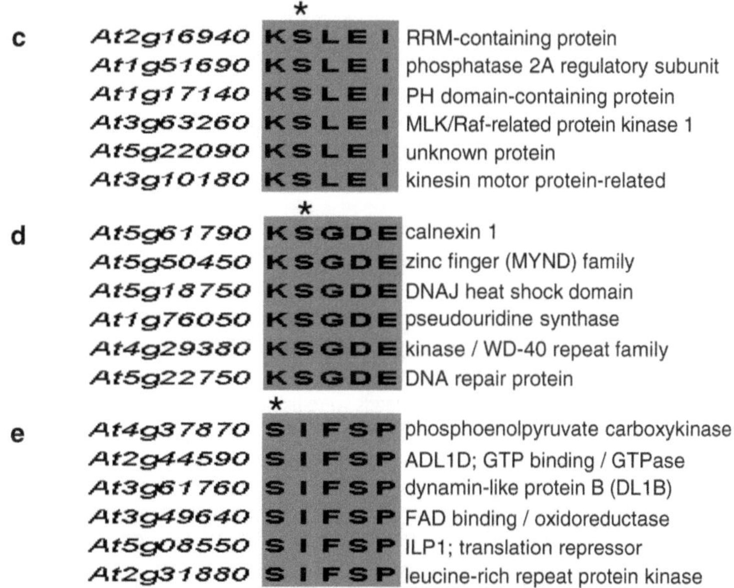

Fig. 1. Bioinformatics analysis of phosphopeptides and construction of phosphorylation motifs using the protein sequence database. (**a**) *T*DDEL, (**b**) RVD*S*S, (**c**) K*S*LEI, (**d**) K*S*GDE, and (**e**) *S*IFSP are five conserved phosphorylation motifs built from both authentic phosphopeptides and protein sequences deposited in the database. The phosphopeptide sequences determined by phosphoproteomic analysis are placed on the top of each group, whereas the rest are the proteins with annotations found in databases. Proteins with unknown functions were omitted (11).

1. Fresh tissue powder of 100 mg is mixed with 300 μl extraction buffer. The mixture is vortexed and incubated on the ice for about 10 min. The cell lysate is then centrifuged at 4°C for 10 min at 14,000 × *g*.

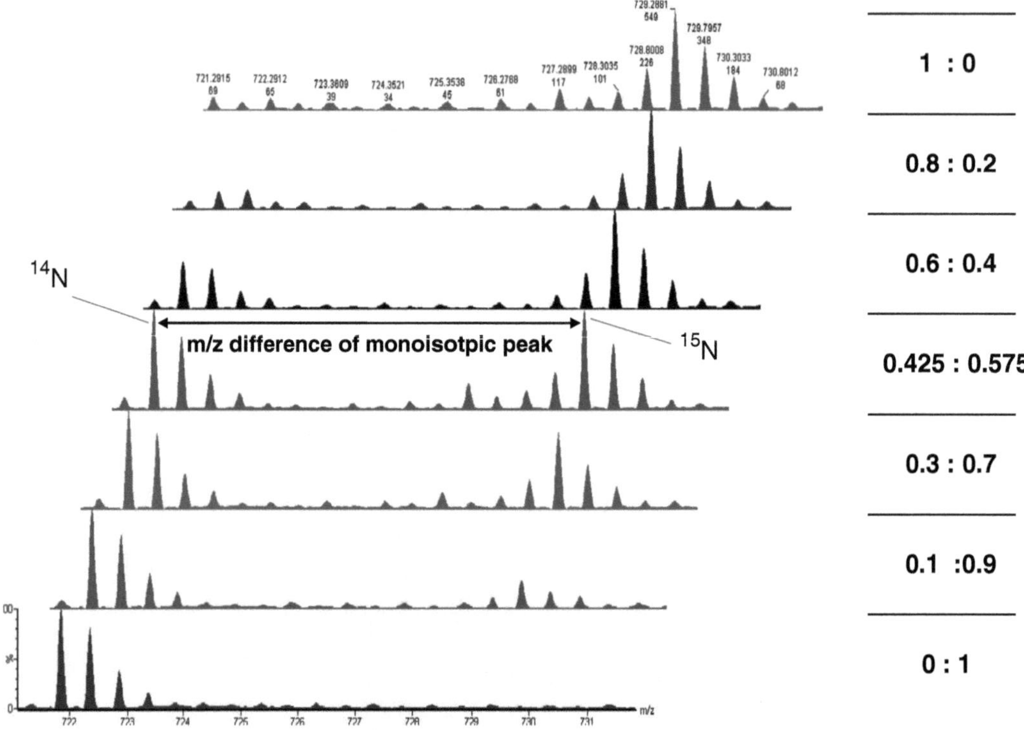

Fig. 2. The mass shift of $^{14}N/^{15}N$-labeled $[M+2H]^{2+}$ precursor ion is 6.9852 *m/z*, corresponding to 13.9704 Da, indicating that the peptide has 14 nitrogen atoms. $^{14}N/^{15}N$-labeled total protein mixtures were extracted from *Arabidopsis thaliana* tissue separately, mixed with different ratios (16:1, 8:1, 2:1, 1:1.3, 1:2, 1:8, and 1:16), and in-solution digested. The resulting peptide was desalted. Fe^{3+}-NTA beads were used to purify phosphopeptides in each sample. The eluting peptides are then subjected to reverse-phase LC–MS/MS analysis. The MS spectrums were acquired by ultra-performance liquid chromatography (nanoAcquity) coupled to a Q-Tof mass spectrometer (micromass, Waters Corporation).

2. Cell lysate of 100 μl is taken out to mix with 25 μl activation mix, 10 μg peptides, and 2.8 μl 50 mM ATP, and incubated at 30°C for 1 h.
3. Ni-NTA beads are firstly equilibrated with extraction buffer. Each sample solution requires 60 μl Ni-NTA slurry. Histidine-tagged substrate peptide is then purified from plant kinase extracts and used for LC–MS/MS analysis. The assayed peptides are purified via an end-over-end rotation for 10 min. Wash with 1 ml trypsin digestion buffer for three times. Add 90 μl trypsin digestion buffer and 1 μg trypsin, and incubate at 37°C for 4 h. Add 10% (*v/v*) acetic acid to terminate the reaction and save the eluate solution.
4. Use Ziptip to remove the salts from the digested sample as stated in Subheading 3.5. The sample is then subject to LC/MS MS analysis. Figure 3 is an example of in vitro kinase assay result (14).

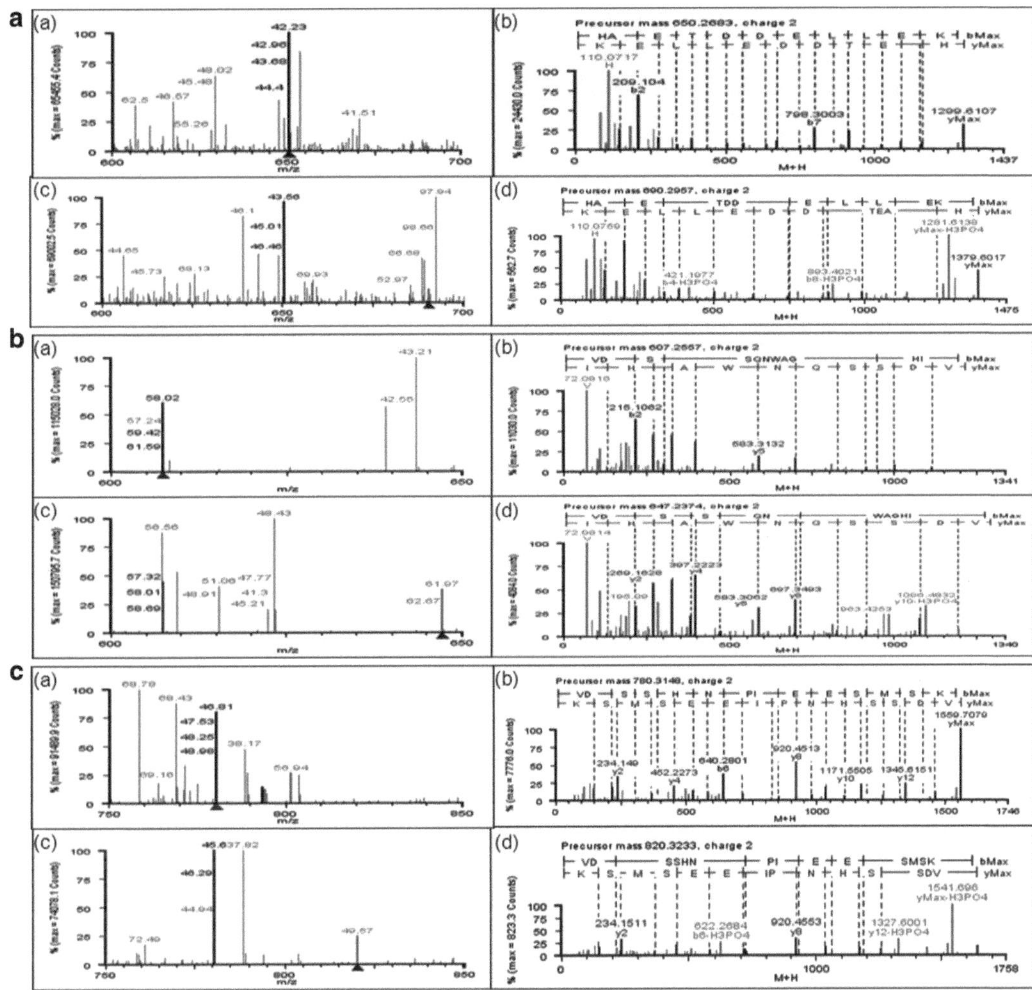

Fig. 3. MS and MS/MS spectra of the phosphorylated peptides produced from the in vitro kinase assays. (**a**) HAETDDELLEK, (**b**) VDSSQNWAGHI, and (**c**) VDSSHNPIEESMSK are three synthetic peptides derived from phosphorylation motifs and used as substrates in the in vitro kinase assays. (**a**, **b**) MS and MS/MS spectra of the substrate peptides, respectively. (**c**, **d**) MS and MS/MS spectra of the *Arabidopsis* kinase-treated synthetic substrate peptides, respectively. Retention time of these phosphopeptides is marked on the top of the mass ion; the precursor ion having the MS/MS spectrum is indicated by an *arrow* underneath, whereas the precursor masses and charges are displayed in the MS/MS spectrum (11).

3.7. Validation of the Functional Role(s) of a Phosphosite in Cell Signaling Using Site-Directed Mutagenesis

Provided that a phosphorylation site has been identified either from MS/MS analysis or bioinformatics-based prediction, the next immediate work is to make point mutations on the phosphosite. Usually, a pair of mutations will be made on the selected phosphosite: S (or T) is mutated to A or I, and Y to F, which is equivalent to a constant dephosphorylation-mimic mutation, and alternatively, S (or T, Y) is mutated to D or E, a constant phosphorylation-mimic mutation. This pair of site-directed mutant genes together with its corresponding wild-type gene are transformed into *Arabidopsis* to examine the in planta function of the phosphorylated protein of interest. Phenotypes of all three transgenic plants that ectopically express mutant proteins and the control wild-type proteins are analyzed and compared to determine the possible roles of phosphorylation site in planta.

4. Notes

1. Too many seeds per jar will make seedlings too crowded to grow well, while too few seeds per jar will make it hard to collect enough tissue for experiment.
2. Although 150 μM AOA is added to the medium and supposed to remove majority of the endogenous ethylene production, there is still trace amount of ethylene produced. This step intends to remove the endogenous ethylene as much as possible so that the effect of ethylene treatment would be more obvious.
3. Pipette up and down slowly to dissolve. Avoid introducing bubbles into the protein sample to save protein from degradation.
4. Prolonging the time of stain will make the dye hard to be removed and, therefore, interfere the downstream steps.
5. Cube size is crucial for protein digestion efficiency and peptide extraction efficiency. Large pieces of gel make the trypsin hard to be taken into the gel and the digested peptides are hard to be extracted out. However, cube smaller than 1 mm^3 causes trouble when exchanging buffers.
6. This step is crucial for efficient enzyme digestion. Wait until all the gel pieces become rehydrated. Add more trypsin solution if some of the gel pieces still not rehydrate.
7. This is to enrich the leftover phosphor peptides in the flow-through fraction of Fe^{3+}-NTA beads, as TiO_2 beads is less sensitive with salt interference during purification process.
8. Sample's pH needs to be lower than 3 for efficient binding to the Ziptip matrix.

Acknowledgments

Thanks Mr. Zhu, Lin and Mr. Guo, and Guang Yu for contributing and editing the text of the protocol. This work is supported by research grants CAS10SC01, 66148, 661207, RPC07/08.SC16, SBI08/09.SC08, GMGS08/09.SC04, and N_HKUST627/06 awarded to Ning Li.

References

1. Krebs EG (1983) Historical perspectives on protein phosphorylation and a classification system for protein kinases. Philos Trans R Soc Lond B Biol Sci 302:3–11
2. Mason MG, Schaller GE (2005) Histidine kinase activity and regualtion of ethylene sginal trasnduction. Can J Bot 83:563–570
3. Ficarro SB, McCleland ML, Stukenberg PT, Burke DJ, Ross MM, Shabanowitz J, Hunt DF, White FM (2002) Phosphoproteome analysis by mass spectrometry and its application to *Saccharomyces cerevisiae*. Nat Biotechnol 20:301–305
4. Nühse TS, Stensballe A, Jensen ON, Peck SC (2003) Large-scale analysis of in vivo phosphorylated membrane proteins by immobilized metal ion affinity chromatography and mass spectrometry. Mol Cell Proteomics 2:1234–1243
5. Nühse TS, Stensballe A, Jensen ON, Peck SC (2004) Phosphoproteomics of the *Arabidopsis* plasma membrane and a new phosphorylation site database. Plant Cell 16:2394–2405
6. Van Bentem S, Anrather D, Roitinger E, Djamei A et al (2006) Phosphoproteomics reveals extensive in vivo phosphorylation of *Arabidopsis* proteins involved in RNA metabolism. Nucleic Acids Res 34:3267–3278
7. Chitteti BR, Peng Z (2007) Proteome and phosphoproteome dynamic change during cell dedifferentiation in *Arabidopsis*. Proteomics 7:1473–1500
8. Li X, Gerber SA, Rudner AD, Beausoleil SA et al (2007) Large-scale phosphorylation analysis of α-factor-arrested *Saccharomyces cerevisiae*. J Proteome Res 6:1190–1197
9. Ono M, Shitashige M et al (2006) Label-free quantitative proteomics using large peptide data sets generated by nanoflow liquid chromatography and mass spectrometry. Mol Cell Proteomics 5:1338
10. Tabata T, Sato T et al (2007) Pseudo internal standard approach for label-free quantitative proteomics. Anal Chem 79:8440–8445
11. Li H, Wong WS, Zhu L, Guo HW, Ecker J, Ning LI (2009) Phosphoproteomics analysis of ethylene-regulated protein phosphorylation in etiolated seedlings of *Arabidopsis* mutant *ein2* using 2-D-separations coupled with a hybrid Q-TOF mass spectrometry. Proteomics 9:1646–1661
12. Gygil S, Rist B, Gerber S, Turecek F, Gelb M, Aebersold R (1999) Quantitative analysis of complex protein mixtures using isotope-coded affinity tags. Nat Biotechnol 17:994–999
13. Yao X, Freas A et al (2001) Proteolytic 18O labeling for comparative proteomics: model studies with two serotypes of adenovirus. Anal Chem 73:2836–2842
14. Ong SE, Blagoev B, Kratchmarova I, Kristensen DB, Steen H, Pandey A, Mann M (2002) Stable isotope labeling by amino acids in cell culture, SILAC, as a simple and accurate approach to expression proteomics. Mol Cell Proteomics 1:376–386
15. Goshe M, Smith R (2003) Stable isotope-coded proteomic mass spectrometry. Curr Opin Biotechnol 14:101–109
16. Whitelegge J, Katz J et al (2004) Subtle modification of isotope ratio proteomics; an integrated strategy for expression proteomics. Phytochemistry 65:1507–1515
17. Ross P, Huang Y et al (2004) Multiplexed protein quantitation in *Saccharomyces cerevisiae* using amine-reactive isobaric tagging reagents. Mol Cell Proteomics 3:1154
18. Huttlin E, Hegeman A et al (2007) Comparison of full versus partial metabolic labeling for quantitative proteomics analysis in *Arabidopsis thaliana*. Mol Cell Proteomics 6:860
19. Liu H, Zhang Y et al (2007) Non-gel-based dual 18O labeling quantitative proteomics strategy. Anal Chem 79:7700–7707
20. Dunkley T, Watson R et al (2004) Localization of organelle proteins by isotope tagging (LOPIT). Mol Cell Proteomics 3:1128
21. Jones A, Bennett M et al (2006) Analysis of the defence phosphoproteome of *Arabidopsis thaliana* using differential mass tagging. Proteomics 6:4155–4165
22. Rudella A, Friso G, Alonso JM, Ecker JR, van Wijk KJ (2006) Downregulation of ClpR2 leads to reduced accumulation of the ClpPRS protease complex and defects in chloroplast biogenesis in *Arabidopsis*. Plant Cell 18:1704–1721
23. Schaff JE, Mbeunkui F, Blackburn K, Bird DM, Goshe MB (2008) SILIP: a novel stable isotope labeling method for in planta quantitative proteomic analysis. Plant J 56:840–854
24. Ong S, Kratchmarova I, Mann M (2003) Properties of 13C-substituted arginine in stable isotope labeling by amino acids in cell culture (SILAC). J Proteome Res 2:173–181
25. Everley P, Krijgsveld J et al (2004) Quantitative cancer proteomics: stable isotope labeling with amino acids in cell culture (SILAC) as a tool for prostate cancer research. Mol Cell Proteomics 3:729

26. Gehrmann M, Hathout Y et al (2004) Evaluation of metabolic labeling for comparative proteomics in breast cancer cells. J Proteome Res 3:1063–1068
27. Rios-Estepa R, Lange B (2007) Experimental and mathematical approaches to modeling plant metabolic networks. Phytochemistry 68:2351–2374
28. Beynon R, Pratt J (2005) Metabolic labeling of proteins for proteomics. Mol Cell Proteomics 4:857
29. Engelsberger W, Erban A, Kopka J, Schulze W (2006) Metabolic labeling of plant cell cultures with K_{15} NO_3 as a tool for quantitative analysis of proteins and metabolites. Plant Methods 2:14
30. Washburn M, Ulaszek R et al (2002) Analysis of quantitative proteomic data generated via multidimensional protein identification technology. Anal Chem 74:1650–1657
31. Krijgsveld J, Ketting R et al (2003) Metabolic labeling of *C. elegans* and *D. melanogaster* for quantitative proteomics. Nat Biotechnol 21:927–931
32. Wu J, Kobayashi M et al (2005) Differential proteomic analysis of bronchoalveolar lavage fluid in asthmatics following segmental antigen challenge. Mol Cell Proteomics 4:1251
33. Nelson C, Huttlin E et al (2007) Implications of 15N-metabolic labeling for automated peptide identification in *Arabidopsis thaliana.* Proteomics 7:1279–1292
34. Hebeler R, Oeljeklaus S et al (2008) Study of early leaf senescence in *Arabidopsis thaliana* by quantitative proteomics using reciprocal 14N/15N labeling and difference gel electrophoresis. Mol Cell Proteomics 7:108
35. Kolkman A, Olsthoorn M et al (2005) Comparative proteome analysis of *Saccharomyces cerevisiae* grown in chemostat cultures limited for glucose or ethanol. Mol Cell Proteomics 4:1
36. Zybailov B, Coleman M et al (2005) Correlation of relative abundance ratios derived from peptide ion chromatograms and spectrum counting for quantitative proteomic analysis using stable isotope labeling. Anal Chem 77:6218–6224
37. Snijders A, de Vos M et al (2005) Novel approach for peptide quantitation and sequencing based on 15N and 13C metabolic labeling. J Proteome Res 4:578–585
38. Guo GY, Li N (2011) Relative and Accurate Measurement of Protein Abundance in (15)N Stable Isotope Labeling in Arabidopsis (SILIA). Phytochemistry. 72: 1028–1039

Chapter 3

Identification of O-linked β-D-*N*-acetylglucosamine-Modified Proteins from *Arabidopsis*

Shou-Ling Xu, Robert J. Chalkley, Zhi-Yong Wang, and Alma L. Burlingame

Abstract

The posttranslational modification of proteins with O-linked β-D-*N*-acetylglucosamine (O-GlcNAc) on serine and threonine residues occurs in all animals and plants. This modification is dynamic and ubiquitous, and regulates many cellular processes, including transcription, signaling and cytokinesis and is associated with several diseases. Cycling of O-GlcNAc is tightly regulated by O-GlcNAc transferase (OGT) and O-GlcNAcase (OGA). Plants have two OGTs, SPINDLY (SPY) and SECRET AGENT (SEC); disruption of both causes embryo lethality. Despite O-GlcNAc modification of proteins being discovered more than 20-years ago, identification and mapping of protein GlcNAcylation is still a challenging task. Here we describe the use of lectin affinity chromatography combined with electron transfer dissociation mass spectrometry to enrich and to detect O-GlcNAc modified peptides from *Arabidopsis*.

Key words: O-GlcNAc, *Arabidopsis*, High-performance liquid chromatography, Mass spectrometry, Electron transfer dissociation, Collision-induced dissociation

1. Introduction

The monosaccharide O-linked β-D-*N*-acetylglucosamine (O-GlcNAc) modification is a ubiquitous and key modification of nuclear and cytoplasmic proteins (1–3). Perturbation of O-GlcNAc levels is associated with many diseases, such as cancer, diabetes, Alzheimer's, and cardiovascular diseases (4–6). Genetic data has also shown that O-GlcNAcylation is critical for embryonic stem cell viability, as both mice and plants show an embryo-lethal phenotype when O-GlcNAc transferase (OGT) functions are disrupted (7, 8). Different from animals, plants have two distinct OGTs, SPY and SEC (7, 9–12).

Zhi-Yong Wang and Zhenbiao Yang (eds.), *Plant Signalling Networks: Methods and Protocols*, Methods in Molecular Biology, vol. 876, DOI 10.1007/978-1-61779-809-2_3,

There is growing evidence that O-GlcNAc and phosphorylation can play reciprocal roles in regulating protein functions and a "yin-yang" model has been proposed for the possible relationship. For instance, the c-Myc oncoprotein is majorly O-GlcNAcylated at Threonine 58, a known site phosphorylated by the kinase, GSK3β, and a mutational hot spot in lymphomas (13). The two PTMs might compete for the modification of the same or proximal Ser/Thr residues (6).

The understanding of O-GlcNAc regulatory functions has been greatly hampered by a lack of knowledge of the identities of the exact residues to which the sugar is attached for most modified proteins. This situation has been due to lack of effective methods for enrichment, detection, and site assignment. However, recent developments in enrichment of either natively modified (14, 15) or tagged/derivatized peptides combined with either electron capture or transfer dissociation mass spectrometry have provided robust modification site analysis tools (16, 17). Mapping the sites of O-GlcNAcylation is critical not only to elucidate the direct function of the modification (by site-directed mutagenesis and/or antibodies) but also to gain a more mechanistic understanding of potential cross talk between GlcNAcylation and other modifications, including phosphorylation.

2. Materials

2.1. Extraction of Total Proteins from Arabidopsis Tissues

1. Prepare stocks separately: 1 M Tris–HCl, pH 8.0; 0.5 M ethylene glycol tetraacetic acid (EGTA), pH 8.0; 0.5 M ethylenedinitrilo tetraacetic acid (EDTA), pH 8.0. Autoclave before storage at room temperature. Prepare stocks: 20% (w/v) sodium dodecyl sulfate (SDS), and store at room temperature.
2. Prepare inhibitor 0.5 mM O-(2-acetamido-2-deoxy-D-glucopyranosylideneamino)*N*-phenylcarbamate (PUGNAc) in water. Make aliquots and store at −20°C (Sigma).
3. Protease inhibitor cocktail (Roche). Store at −20°C.
4. Liquid nitrogen.
5. Mortar and pestle.
6. Extraction buffer Y: 100 mM Tris–HCl, pH 8.0, 2% SDS, 1% β-mercaptoethanol, 5 mM EGTA, 10 mM EDTA, 20 μM PUGNAc, and 1× protease inhibitor cocktail. Make freshly each time by mixing aliquots from stocks.
7. Phenol (Tris buffered, pH 7.5–7.9). Store at 4°C.
8. Extraction buffer Z: 50 mM Tris–HCl, pH 8.0. Store at 4°C.
9. Methanol. Store at 4°C.

10. 0.1 M ammonium acetate in methanol. Store at 4°C.
11. Lysis buffer: 6 M guanidine-HCl. Store at room temperature.
12. Biorad protein assay kit (Bio-Rad).

2.2. Tryptic Digestion of Protein Samples

1. Prepare 50 mM NH_4HCO_3 in water. Store at room temperature.
2. Reduction reagent: 1 M Tris (2-carboxyethyl) phosphine (TCEP) in 50 mM NH_4HCO_3. Make aliquots and store in −20°C. Keep in dark.
3. Alkylation reagent: 550 mM iodoacetamide in 50 mM NH_4HCO_3. Make it fresh. Keep in dark.
4. Modified trypsin (Promega).

2.3. Desalting the Peptide Sample by Using C18 Filled Sep-Paks

1. Formic acid.
2. Buffer A: 0.1% formic acid.
3. Buffer B: 70% acetonitrile and 0.1% formic acid.
4. C18 filled Sep-Pak (Waters).
5. Syringes (10 ml).
6. Needles.
7. Speed vacuum.
8. Nitrogen.

2.4. Packing a Lectin Weak Affinity Chromatography Column

1. Resin: WGA-Agarose (Vector Laboratories).
2. Tubing: Teflon PFA 1/16″ (1.6 mm) OD, 1/25″ (1 mm) ID (Upchurch Scientific).
3. Packing a lectin weak affinity chromatography (LWAC) buffer: 25 mM Tris, pH 7.8, 200 mM NaCl, 5 mM $CaCl_2$, and 1 mM $MgCl_2$.
4. LWAC elute buffer: 20 mM GlcNAc (*N*-acetyl-D-glucosamine) (Sigma).
5. Large empty column to use as a reservoir for packing resin from: e.g., AP mini-column 10 mm × 120 mm (Waters).
6. Frit: 2 μm porosity Stainless Steel Frit 0.062″ OD (Upchurch Scientific).
7. Unions for each end of column: e.g., Stainless Steel ZDV Union 0.02″ Thru-hole (Upchurch Scientific).

2.5. Enriching O-GlcNAc-Modified Peptide by Using LWAC Column

1. LWAC buffer: 25 mM Tris, pH 7.8, 200 mM NaCl, 5 mM $CaCl_2$, and 1 mM $MgCl_2$.
2. LWAC elute buffer: 20 mM GlcNAc in LWAC buffer.
3. AKTA purifier HPLC (GE Healthcare).

2.6. Desalting Peptides by Using C18 Pipette Tip

1. Buffer A: 0.1% formic acid.
2. Buffer B: 70% acetonitrile and 0.1% formic acid.
3. C18 100 μl OMIX pipette tips for micro extraction (Varian).

2.7. Detection by Liquid Chromatography–Tandem Mass Spectrometry

1. HPLC: Waters Nanoacquity-Ultra Performance LC (Waters).
2. Linear Ion Trap (LTQ)-Orbitrap XL with Electron Transfer Dissociation (ETD) (Thermo).
3. HPLC Solvent A: Water/0.1% formic acid.
4. HPLC Solvent B: Acetonitrile/0.1% formic acid.
5. Trapping column: 5 μm Symmetry C18, 180 μm inner diameter ×20 mm (Waters).
6. Analytical column: 1.7 μm BEH130 C18 100 μM inner diameter ×100 mm (Waters).

3. Methods

The lectin wheat germ agglutinin (WGA) has affinity for terminal *N*-acetylglucosamine (GlcNAc) and sialic acid residues. It has four binding sites, so can bind with high-affinity to branched glycan structures with multiple terminal GlcNAc moieties. However, the affinity for a single GlcNAc residue, as encountered with the regulatory modification of O-GlcNAcylation, found on serines and threonines of nuclear and cytoplasmic proteins, is low. An LWAC protocol has been developed to enrich for O-GlcNAc-modified peptides (14), in which WGA attached to agarose is packed into a long column. Modified peptides can be separated from unmodified peptides through their retardation on these long columns during isocratic HPLC, causing them to elute later than unmodified peptides.

O-GlcNAcylation site identification using conventional collision-induced dissociation (CID) analysis in a mass spectrometer is usually not possible. CID is a vibrational activation fragmentation process that breaks the weakest bonds in the structure. The O-glycosidic link is significantly more labile than the peptide backbone. Hence, the O-GlcNAc moiety is readily liberated as an oxonium ion before peptide backbone fragmentation, and site assignment cannot be derived from the resulting deglycosylated fragment ions. The recently developed electron capture dissociation and ETD are a radical-based “non-ergodic” fragmentation process that results in mostly peptide backbone cleavage, thus enabling the identification of sites bearing labile posttranslational modifications (14, 15, 18).

3.1. Extraction of Total Proteins from Arabidopsis Tissues

1. Harvest flower tissues from 45-day-old *Arabidopsis* plant growing in green house. Freeze in liquid nitrogen.
2. Grind tissues in a mortar with liquid nitrogen to a fine powder and weigh 2 g tissue powder in 50-ml tube.
3. Add three volume (6 ml) of buffer Y (see Notes 1 and 2). Vortex for 1 min.
4. Heat the samples for 10 min at 65°C.
5. Centrifuge at 20,000 × *g* for 20 min at 20°C.
6. Transfer the supernatant to new tubes. Add equal volume of ice-cold phenol (Tris buffer, pH 7.5–7.9), and vortex for 1 min.
7. Centrifuge at 20,000 × *g* for 15 min at 4°C to separate phenol and aqueous phases.
8. Remove the upper aqueous phase to leave the interface intact. Discard the upper aqueous phase.
9. Re-extract the phenol phase twice with ice-cold Buffer Z as in steps 7 and 8.
10. Mix with five volumes of cold 0.1 M ammonium acetate in methanol and leave at –20°C overnight to precipitate proteins.
11. Centrifuge at 20,000 × *g* for 20 min at 4°C. Keep the pellet. Remove all the supernatant.
12. Wash the protein pellet twice with 1 ml ice-cold 0.1 M ammonium acetate in methanol and 1 ml cold methanol twice; centrifuge for 5 min and remove the liquid each time (see Note 3).
13. Resuspend the protein pellet in lysis buffer.
14. Centrifuge at 20,000 × *g* for 20 min.
15. Transfer the supernatant to a new tube, and determine the protein concentration with Bio-Rad protein assay kit using BSA as a standard.

3.2. Tryptic Digestion of Protein Samples

1. Reduce the disulfide bonds of 2 mg protein sample by adding Tris TCEP to final concentration of 2 mM for 60 min at 56°C.
2. Alkylate the free cysteines of the protein sample by adding iodoacetamide to final concentration of 10 mM for 60 min at room temperature in the dark.
3. Dilute the sample with 50 mM NH_4HCO_3 to make the final guanidine-HCl concentration of 1.5 M (see Notes 4–6).
4. Add modified trypsin 1:50 w/w overnight at 37°C.
5. Quench the protease activity by acidification of the reaction mixture with formic acid to final concentration of 1% formic acid.
6. Centrifuge at 20,000 × *g* for 10 min to remove insoluble material. Keep the supernatant.

3.3. Desalting the Peptide Sample by Using C18 Sep-Pak

1. Cut Sep-Pak column ends to reduce dead volume.
2. Activate the C18 Sep-Pak column: Attach a needle to the syringe and pull buffer B (2 ml); detach the needle, and attach the syringe to the Sep-Paks; and slowly push the buffer through the Sep-Pak. When the buffer has been pushed through the Sep-Pak, remove the syringe from the Sep-Pak and leave Sep-Pak on clean tissue paper.
3. Equilibrate the C18 Sep-Pak column: Attach another needle to the syringe and pull buffer A (8 ml); detach the needle, and attach the syringe to the Sep-Paks; and slowly push the buffer through the Sep-Pak.
4. Repeat this equilibration step twice to remove any trace of acetonitrile (see Note 7).
5. Load the C18 Sep-Pak column with the peptide sample: Attach another needle to the syringe and pull peptide samples; detach the needle and attach the syringe to the Sep-Paks; slowly push the buffer through the Sep-Pak, collecting the flow through back into the original peptide sample tube.
6. Repeat the loading step four more times to ensure maximal binding of the peptides to the column.
7. Wash the Sep-Pak column to remove salts: Attach another needle to a new syringe and pull buffer A 8 ml; detach the needle and attach the syringe to the Sep-Paks; and slowly push the buffer through the Sep-Pak, discard the flow through.
8. Repeat the washing step four more times to completely wash away salts and other contaminants.
9. Elute the peptides from the Sep-Pak: Attach another needle to a new syringe and pull buffer B 1 ml; detach the needle and attach the syringe to the Sep-Paks; and slowly push the buffer through the Sep-Pak, and collect the eluate.
10. Repeat the elution step and collect the eluate in the same tube.
11. Dry the tube in Speedvac to complete dryness.
12. Store the peptides in −80°C freezer before use.

3.4. Packing a Lectin Weak Affinity Chromatography Column

1. Insert Frit into union and attach to one end of Teflon tubing (see Note 8).
2. Attach reservoir column to the other end of Teflon tubing (see Note 9).
3. Partially fill reservoir column with WGA-agarose resin.
4. Pack resin into Teflon tubing using an HPLC pump delivering LWAC buffer at a flow rate of 50–200 μl/min, making sure that the back pressure never exceeds 2 MPa (20 mbar) (see Notes 10 and 11).

5. It may be necessary to replenish the reservoir column with more resin on several occasions (see Note 12). When doing this, stop the HPLC and wait for pressure to drop to zero before disconnecting.
6. When the column is packed to a long enough length (see Note 13), stop the packing, and then cut the back end of the column at the point at which it is packed up to in order to remove dead volume.
7. Attach union to the back end of the column, and then cap each end to prevent column from drying out.
8. Store the LWAC column at 4°C.

3.5. Enriching O-GlcNAc-Modified Peptides by Using LWAC Column

1. Resuspend the peptides in 200 μl LWAC buffer.
2. Install the LWAC column and flush lines with LWAC buffer for 5 min at a flow rate of 100 μl/min.
3. Load sample and wait for the main peak to elute.
4. Manually collect fractions at 1-min intervals during the elution of the main UV-visible peak and subsequent tail of the peak (see Note 14). An example of the HPLC results produced is shown in Fig. 1. A total of more than ten fractions are analyzed by mass spectrometry to identify GlcNAc-modified peptides.
5. At the end of the run, inject 200 μl of LWAC buffer containing 20 mM GlcNAc to elute any complex glycans (see Note 15).
6. Store the column at 4°C.
7. Store the fractions at −80°C freezer before further use.

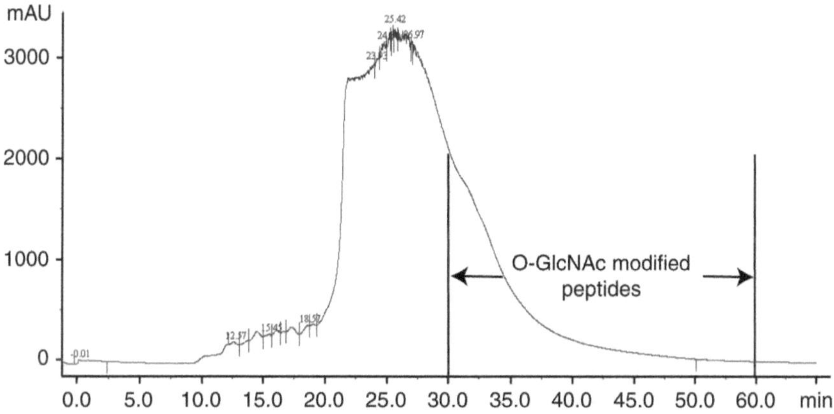

Fig. 1. Lectin weak affinity chromatography enrichment of *Arabidopsis* O-linked β-D-*N*-acetylglucosamine (O-GlcNAc)-modified peptides using a WGA-Agarose column with LWAC buffer at a flow rate of 100 μl/min. Peptide elution is monitored at 205 nm. The region containing (O-GlcNAc) modified peptides is labeled.

3.6. Desalting Peptides by Using C18 Pipette Tip

1. Sample treatment: Adjust sample to a 1% formic acid concentration using a 10% formic acid solution.
2. Activate tip: Aspirate 100 μl buffer B and discard solvent.
3. Equilibration: Aspirate 100 μl buffer A and discard solvent. Repeat four more times to remove any trace of acetonitrile (see Note 7).
4. Apply sample: Aspirate and dispense pretreated sample into the tip three to five times (see Notes 16 and 17).
5. Wash tip: Aspirate buffer A and discard solvent. Repeat.
6. Elute peptides: Aspirate 100 μl buffer B and pipette the tip five times.
7. Reduce the sample volume and remove the acetonitrile by vacuum concentration.
8. Store the peptides at −80°C freezer or directly resuspend the peptides in buffer A for further mass spectrometry analysis.

3.7. Detection by Liquid Chromatography–Tandem Mass Spectrometry

1. The peptide mixtures are analyzed by online nanoflow liquid chromatography–tandem mass spectrometry (LC–MS/MS) on a Nanoacquity ultraperformance LC system connected to an LTQ-XL Orbitrap with ETD source.
2. Peptides are first loaded onto a trapping column packed with C18, and then washed with 0.1% formic acid. The trapping column is connected to an analytical column.
3. The reverse-phase liquid chromatography is performed at a flow rate of 400 nl/min.
4. A 90-min gradient is used; peptides are eluted by a gradient from 2 to 30% solvent B over 65 min followed by a short wash at 50% solvent B, before returning to starting conditions. Peptide components elute over a period of ~65 min during these runs.
5. The effluent from the HPLC column is directly interfaced with the mass spectrometer.
6. The LTQ Orbitrap XL instrument is operated in data-dependent mode to automatically switch between full-scan MS and MS/MS acquisition.
7. After a precursor scan of intact peptides is measured in the orbitrap by scanning from *m/z* 350–1,400, the three most intense multiply charged precursors are selected for both CID and ETD analysis in the linear ion trap.
8. Activation times are 30 and 100 ms for CID and ETD fragmentation, respectively. Automatic gain control (AGC) targets are 100,000 ions for orbitrap scans and 10,000 for MS/MS scans, and the AGC for the fluoranthene ions used for ETD is 100,000. Supplemental activation of the charge-reduced

species is used in the ETD analysis to improve fragmentation. Dynamic exclusion for 60 s is used to prevent repeated analysis of the same components.

3.8. Data Analysis

1. Separate ETD and CID peak lists are generated from raw data files using an in-house script PAVA, and then CID and ETD data are searched separately using Protein Prospector version 5.7.1 (19, 20) against a database that consists of the UniProtKB protein database downloaded on 1 November 2011, to which a randomized version has been concatenated. Only *Arabidopsis* entries are searched, meaning a total of 101,922 entries are queried. ETD peptide results are reported using a peptide false discovery rate level of 0.5% according to concatenated database search results.
2. For both CID and ETD data, a precursor mass tolerance of 20 ppm and a fragment mass error tolerance of 0.6 Da are allowed.
3. Modification parameters: Carbamidomethylcysteine is searched as a constant modification. Variable modifications include protein N-terminal acetylation, peptide N-terminal glutamine conversion to pyroglutamate, and methionine oxidation. For ETD data, HexNAc modification of Serine or Threonine and asparagine residues is considered. For CID data, HexNAc modification of serine or threonine and asparagines residues is considered, but in addition, a modification observed as a neutral loss of 203.08 Da (so all fragments are assumed not to have GlcNAc attached) is considered from serines and threonines (see Note 18).
4. Assignments of all modified peptides are checked manually; in every case, the CID results are consistent with the ETD assignment (see Note 18). An example result is shown in Fig. 2.

4. Notes

1. This extraction method is adapted from the protocol for preparation of proteins for 2-D DIGE (21). Alternative approaches to extract proteins would be acceptable as long as O-GlcNAc inhibitors are included in the extraction buffer.
2. The O-GlcNAc modification can be easily removed by hydrolases, such as O-GlcNAcase or hexosaminidases during cell lysis. Hence, it is important to include inhibitors during the extraction and purification processes that preserve the levels of O-GlcNAc on proteins. PUGNAc is a potent competitive inhibitor of hexosaminidases that inhibits all three known human hexosamindases [HEXA/B and O-GlcNAcase (OGA)] (22).

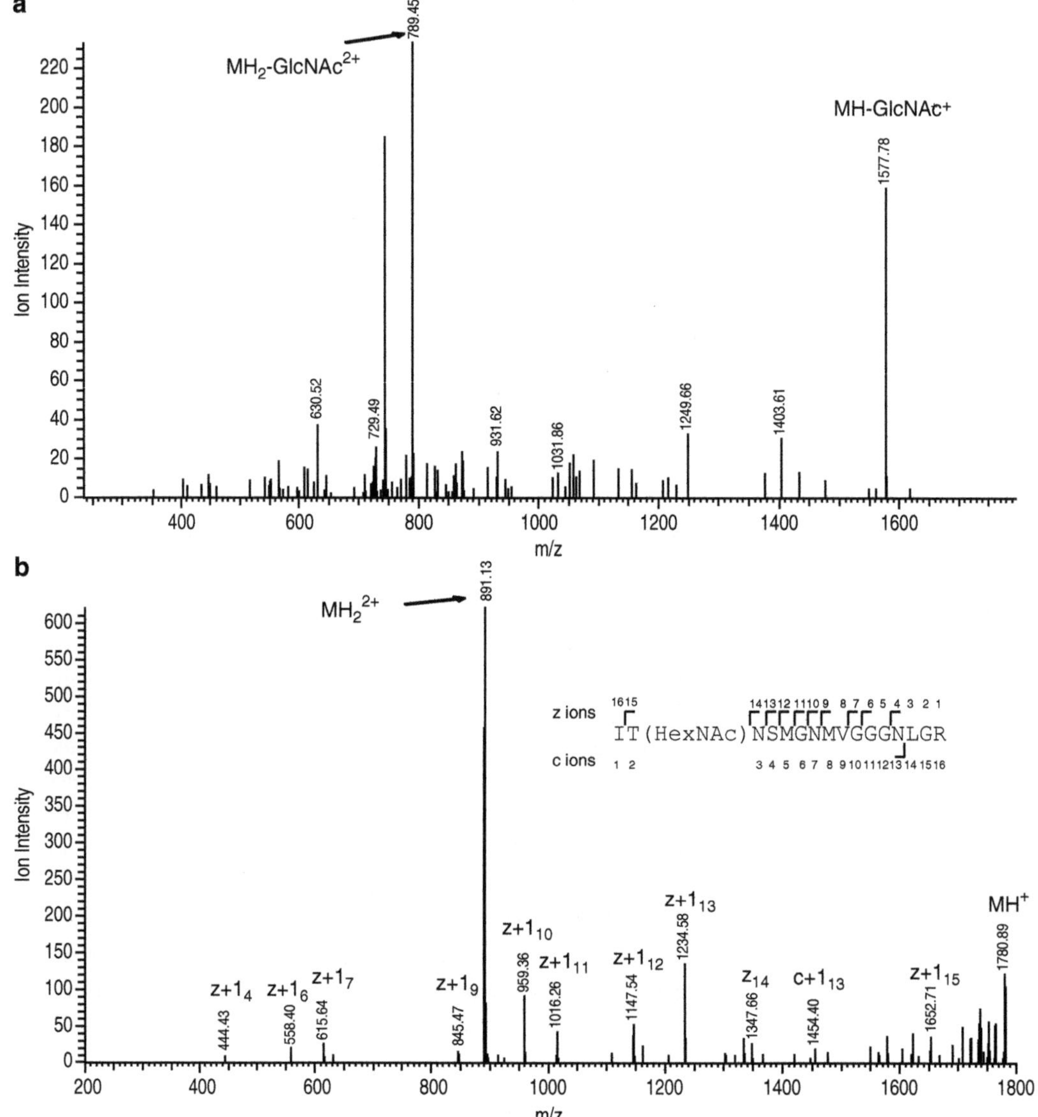

Fig. 2. Collision-induced dissociation (CID) and electron transfer dissociation (ETD) spectra of an *m/z* 890.9212 2+ precursor (theoretical *m/z* = 890.9195, +2.0 ppm). (**a**) The molecular weight of this peptide is shifted to higher mass of 203 Da and the CID spectrum displays loss of sugar [MH_2-GlcNAc^{2+}] and [MH-GlcNAc^{+}], showing this peptide to be O-linked β-D-*N*-acetylglucosamine (O-GlcNAc) modified. (**b**) Interpretation of the ETD spectrum of the species of *m/z* 890.9212 establishes a doubly charged peptide ITNSMGNMVGGGNLGR with Threonine 2 being modified by O-GlcNAc.

3. It is better not to disturb the pellets at this stage. Remove all the liquid and let the pellet dry for 1 min.
4. Maximum allowable concentration of guanidine-HCl for trypsin is less than 1.5 M.
5. To achieve the highest digestion efficiency, optimize the solution close to pH 7.5–9.0.

6. It is acceptable to use urea with or without thiourea to dissolve the protein sample; however, long incubation in the presence of urea may lead to carbamylation of the available amino terminus (43 Da mass increase) of the proteins as well as the side chain (ε-amino groups) of lysine residues (43 Da mass increase; resulting in a protein that is unsuitable for many enzymatic digests including tryptic cleavage). Maximum allowable concentration of urea and thiourea for trypsin is less than 1 M total.
7. Binding of peptides to the C18 (this is an 18 carbon hydrocarbon chain that is bonded to the silicate) matrix is a result of hydrophobic interactions. For efficient binding, sample solutions must be free from any organic solvent because even a small proportion of organic solvent can prevent adsorption of some peptides.
8. A frit is used in the in-line filter to remove unwanted particulate from the solution.
9. To get a good separation of modified peptides from the unmodified peptides, we recommend packing a column of at least 3 meters in length (we have used columns up to 10 meters in length).
10. An issue with these columns is that the agarose can compress. Hence, we recommend keeping the column pressure below 2 MPa (20 mbar). This means operating at a flow rate between 50–150 μl/min (longer columns require lower flow rates).
11. On occasions, the column may stop packing. If this happens, a possible solution may be to agitate the reservoir column. If this still does not remedy the problem, then stop the HPLC (wait for the pressure to drop to zero). It is then possible to reverse the direction of the column by moving the frit to the other end of the column.
12. Packing the tubing requires 0.8 ml of resin per meter of column.
13. Packing these columns is slow—expect it to take all day to make a column.
14. Depending on the length of your column, the O-GlcNAc-modified peptides elute in the tail of the main peak or a little afterwards. Do not worry if you do not see a distinct peak for modified peptides, and collect fractions even beyond the point at which there is any obvious UV absorbance (there will probably still be peptides there).
15. Inject LWAC Elute Buffer to elute any complex glycans that may be bound to the column. The GlcNAc in this buffer absorbs at 214 nm, so do not be concerned if you observe a large UV peak.
16. To achieve maximum sample binding, the whole sample must be passed several times through the C18 matrix.

17. For steps 4–6, to achieve optimum flow and peptide recovery, do not introduce air through the tip.

18. Modified peptides spectra need to be manually inspected, particularly for site assignment reliability. One characteristic of O-linked GlcNAc-modified peptides is the loss of O-GlcNAc in CID spectra. N-linked GlcNAc-modified peptides have been detected in *Arabidopsis* O-GlcNAc-enriched samples, similar as reported in ref 15. However, N-linked GlcNAc is stable in CID mass spectrometry, so loss of GlcNAc will not be detected from N-linked GlcNAc-modified peptides.

Acknowledgments

This work was financially supported by the Division of Chemical Sciences, Geosciences, and Biosciences, Office of Basic Energy Sciences of the US Department of Energy through Grant DE-FG02-08ER15973, and by NIH (R01GM066258) and NSF (IOS-0724688). The UCSF Mass Spectrometry Facility (A.L.B., Director) is supported by the Biomedical Research Technology Program of the National Centre for Research Resources, NIH NCRR RR01614.

References

1. Hart GW, Slawson C, Ramirez-Correa G, Lagerlof O (2011) Cross talk between O-GlcNAcylation and phosphorylation: roles in signaling, transcription, and chronic disease. Annu Rev Biochem 80:6.1–6.34
2. Hart G, Housley M, Slawson C (2007) Cycling of O-linked beta-*N*-acetylglucosamine on nucleocytoplasmic proteins. Nature 446:1017–1022
3. Olszewski NE, West CM, Sassi SO, Hartweck LM (2009) O-GlcNAc protein modification in plants: evolution and function. Biochim Biophys Acta 1800:49–56
4. Yang X, Ongusaha PP, Miles PD, Havstad JC, Zhang F, So WV, Kudlow JE, Michell RH, Olefsky JM, Field SJ, Evans RM (2008) Phosphoinositide signalling links O-GlcNAc transferase to insulin resistance. Nature 451:964–969
5. Ngoh GA, Facundo HT, Zafir A, Jones SP (2010) O-GlcNAc signaling in the cardiovascular system. Circ Res 107:171–185
6. Ozcan S, Andrali SS, Cantrell JE (2010) Modulation of transcription factor function by O-GlcNAc modification. Biochim Biophys Acta 1799:353–364
7. Hartweck LM, Scott CL, Olszewski NE (2002) Two O-linked N-acetylglucosamine transferase genes of *Arabidopsis thaliana* L. Heynh. have overlapping functions necessary for gamete and seed development. Genetics 161:1279–1291
8. Shafi R, Iyer SP, Ellies LG, O'Donnell N, Marek KW, Chui D, Hart GW, Marth JD (2000) The O-GlcNAc transferase gene resides on the X chromosome and is essential for embryonic stem cell viability and mouse ontogeny. Proc Natl Acad Sci USA 97:5735–5739
9. Jacobsen SE, Olszewski NE (1993) Mutations at the SPINDLY locus of *Arabidopsis* alter gibberellin signal transduction. Plant Cell 5:887–896
10. Jacobsen SE, Binkowski KA, Olszewski NE (1996) SPINDLY, a tetratricopeptide repeat protein involved in gibberellin signal transduction in *Arabidopsis*. Proc Natl Acad Sci USA 93:9292–9296
11. Izhaki A, Swain SM, Tseng TS, Borochov A, Olszewski NE, Weiss D (2001) The role of SPY and its TPR domain in the regulation of gibberellin action throughout the life cycle of Petunia hybrida plants. Plant J 28:181–190
12. Hartweck LM, Genger RK, Grey WM, Olszewski NE (2006) SECRET AGENT and

SPINDLY have overlapping roles in the development of *Arabidopsis thaliana* L. Heyn. J Exp Bot 57:865–875

13. Chou TY, Hart GW, Dang CV (1995) c-Myc is glycosylated at threonine 58, a known phosphorylation site and a mutational hot spot in lymphomas. J Biol Chem 270:18961–18965
14. Vosseller K, Trinidad JC, Chalkley RJ, Specht CG, Thalhammer A, Lynn AJ, Snedecor JO, Guan S, Medzihradszky KF, Maltby DA, Schoepfer R, Burlingame AL (2006) O-linked *N*-acetylglucosamine proteomics of postsynaptic density preparations using lectin weak affinity chromatography and mass spectrometry. Mol Cell Proteomics 5:923–934
15. Chalkley RJ, Thalhammer A, Schoepfer R, Burlingame AL (2009) Identification of protein O-GlcNAcylation sites using electron transfer dissociation mass spectrometry on native peptides. Proc Natl Acad Sci USA 106:8894–8899
16. Khidekel N, Ficarro SB, Peters EC, Hsieh-Wilson LC (2004) Exploring the O-GlcNAc proteome: direct identification of O-GlcNAc-modified proteins from the brain. Proc Natl Acad Sci USA 101:13132–13137
17. Wang Z, Udeshi ND, O'Malley M, Shabanowitz J, Hunt DF, Hart GW (2009) Enrichment and site mapping of O-linked *N*-acetylglucosamine by a combination of chemical/enzymatic tagging, photochemical cleavage, and electron transfer dissociation mass spectrometry. Mol Cell Proteomics 9:153–160
18. Kelleher NL, Zubarev RA, Bush K, Furie B, Furie BC, McLafferty FW, Walsh CT (1999) Localization of labile posttranslational modifications by electron capture dissociation: the case of gamma-carboxyglutamic acid. Anal Chem 71:4250–4253
19. Chalkley RJ, Baker PR, Hansen KC, Medzihradszky KF, Allen NP, Rexach M, Burlingame AL (2005) Comprehensive analysis of a multidimensional liquid chromatography mass spectrometry dataset acquired on a quadrupole selecting, quadrupole collision cell, time-of-flight mass spectrometer: I. How much of the data is theoretically interpretable by search engines? Mol Cell Proteomics 4:1189–1193
20. Chalkley RJ, Baker PR, Huang L, Hansen KC, Allen NP, Rexach M, Burlingame AL (2005) Comprehensive analysis of a multidimensional liquid chromatography mass spectrometry dataset acquired on a quadrupole selecting, quadrupole collision cell, time-of-flight mass spectrometer: II. New developments in protein prospector allow for reliable and comprehensive automatic analysis of large datasets. Mol Cell Proteomics 4:1194–1204
21. Deng Z, Zhang X, Tang W, Oses-Prieto JA, Suzuki N, Gendron JM, Chen H, Guan S, Chalkley RJ, Peterman TK, Burlingame AL, Wang ZY (2007) A proteomics study of brassinosteroid response in *Arabidopsis*. Mol Cell Proteomics 6:2058–2071
22. Haltiwanger RS, Grove K, Philipsberg GA (1998) Modulation of O-linked *N*-acetylglucosamine levels on nuclear and cytoplasmic proteins in vivo using the peptide O-GlcNAc-beta-*N*-acetylglucosaminidase inhibitor O-(2-acetamido-2-deoxy-D-glucopyranosylidene) amino-*N*-phenylcarbamate. J Biol Chem 273:3611–3617

Chapter 4

Quantitative Analysis of Protein Phosphorylation Using Two-Dimensional Difference Gel Electrophoresis

Zhiping Deng, Shuolei Bu, and Zhi-Yong Wang

Abstract

Posttranslational modifications of proteins, especially phosphorylation and dephosphorylation, play an important role in signal transduction and cellular regulation in plants. Both 2-DE gel-based and non-gel-based proteomic technologies can monitor the changes in phosphorylation state of proteins. In this chapter, we describe two protocols for discovery and validation of differential protein phosphorylation using affinity enrichment of phosphoproteins by immobilized metal affinity chromatography (IMAC) or protein immunoprecipitation (IP) followed by two-dimensional difference gel electrophoresis (2-D DIGE). We name these methods IMAC-DIGE and IP-DIGE. For IMAC-DIGE, phosphoproteins are enriched from tissue extract using $GaCl_3$-based IMAC and then analyzed by 2-D DIGE, which reveals changes of protein phosphorylation as protein spot shifts. IMAC enrichment improves detection of low-abundance regulatory phosphoproteins. For IP-DIGE, proteins of interest can be immunopurified and then analyzed by 2-D DIGE to confirm changes of posttranslational modifications that alter the charge or size of the proteins.

Key words: Phosphoprotein, Immobilized metal affinity chromatography, Proteomics, 2-D DIGE, Phosphorylation, Posttranslational modification, Brassinosteroid, BZR1

1. Introduction

Posttranslational modifications, such as phosphorylation, acetylation, and ubiquitination, are major mechanisms for regulating protein activity, localization, accumulation, and interaction with other proteins (1). Protein phosphorylation by kinases and dephosphorylation by phosphatases are the most widely used mechanisms for signal transduction and cellular regulation (2). Changes of protein phosphorylation are often the bases for responses to diverse endogenous and environmental signals. Quantitative analysis of protein phosphorylation changes associated with biological responses is a powerful approach to identifying key components and uncovering critical mechanisms underlying cellular responses to various signals (2).

Zhi-Yong Wang and Zhenbiao Yang (eds.), *Plant Signalling Networks: Methods and Protocols*,
Methods in Molecular Biology, vol. 876, DOI 10.1007/978-1-61779-809-2_4,

With the advance in proteomics technology, it is possible to globally profile changes in the abundance or phosphorylation state of proteins and even specific phosphorylation sites. Quantitative analysis at the proteomics level usually involves 2-DE gel-based or non-gel-based approach. In the non-gel-based method, protein mixtures are first digested into peptides, usually by trypsin, and then analyzed by liquid chromatography coupled with tandem mass spectrometry (LC–MS/MS). Quantitative comparison is commonly performed by labeling peptides with isotopic or isobaric mass tags, such as isobaric tag for relative and absolute quantitation (iTRAQ) (3), stable isotope labeling with amino acids in cell culture (SILAC) (4), and isotope-coded affinity tags (ICAT) (5). Quantitative phosphoproteomics has opened new perspectives in delineating signaling pathway (6). One example of such MS-based quantitative phosphoproteomics analysis in plants, using iTRAQ labeling and enrichment of phosphopeptides of plasma membrane proteins (7), identified protein phosphorylation changes induced by pathogen signals in *Arabidopsis thaliana*. One disadvantage of LC–MS/MS approach is that only partial peptide sequence is detected for most proteins.

In contrast to LC–MS/MS, 2-DE gels separate whole proteins based on their charge by isoelectric focusing (IEF) in the first dimension and based on their molecular mass by SDS-PAGE in the second dimension (8). As such, a change of phosphorylation at any residue of the protein can be detected as spot shift along the IEF dimension. In traditional 2-DE, protein spots are visualized and quantified after staining the gels with silver nitrate or fluorescent dyes, such as Sypro Ruby (Invitrogen) or Deep Purple (GE Healthcare). A major drawback in traditional 2-DE is the gel-to-gel variability. Two-dimensional difference gel electrophoresis (2-D DIGE) (9, 10), commercially introduced by GE Healthcare, greatly improves the sensitivity and reproducibility of 2-DE. In 2-D DIGE, two or three protein samples are covalently labeled with different fluorescence dyes and then mixed and separated in the same 2-DE gel. This allows direct comparison of two or three samples in the same gel. Non-gel-based LC–MS/MS and 2-D DIGE approaches are complementary and each has its own limitations and advantages (11–13).

Compared to non-gel-based methods, 2-D DIGE directly analyzes whole proteins without tryptic digestion. In the LC–MS/MS approach, tryptic digestion increases sample complexity and masks some posttranslational modifications, such as proteolytic cleavage; a change of phosphorylation can only be detected if the phosphopeptide is quantified. Thus, a major advantage of 2-D DIGE is that the change of protein modification is detected at the whole protein level, and the site of modification in the protein of interest can be determined in subsequent MS/MS analysis. The disadvantages of 2-DE include its ineffectiveness in resolving proteins with extreme molecular weight or isoelectric points, or hydrophobic proteins. However, neither 2-D DIGE nor LC–MS/MS can detect all proteins of a complex biological sample. Usually, several thousands of proteins can be analyzed, and sample prefractionation is required for detecting low-abundance proteins by either method. 2-D DIGE has been successfully used to study signal transduction in both animals and plants (14–17).

Quantitative phosphoproteomics is challenging largely because many phosphoproteins, especially those involved in signal transduction, are of low abundance and difficult to detect by 2-DE gels or mass spectrometry. Furthermore, the stoichiometry of many phosphopeptides is low, making them difficult to detect. Therefore, enrichment of phosphoproteins or phosphopeptides is often required for detecting phosphorylation of signaling proteins. For example, the change of phosphorylation status of a transcription factor induced by brassinosteroid was not detected in total protein (17), but observed in immobilized metal affinity chromatography (IMAC)-enriched fractions (15).

A number of strategies have been developed for enrichment of phosphoproteins or phosphopeptides, such as affinity purification with phospho-specific antibodies, chemical modification followed by biotin tagging, strong cation exchange chromatography, and IMAC (6). IMAC is the most commonly used approach for phosphoprotein enrichment. In this chapter, we describe a protocol for enrichment of phosphoproteins from *Arabidopsis* tissues by $GaCl_3$-based IMAC, and subsequent analysis of changes in phosphoproteins by 2-D DIGE. This protocol has successfully detected brassinosteroid-induced changes of protein phosphorylation (15, 17). In addition, we provide a method of immunoprecipitation followed by 2-D DIGE for analyzing changes of phosphorylation or other modifications of individual proteins of interest.

2. Materials

2.1. Preparation of Total Protein Samples

1. Tris–HCl buffer (30×): 1.5 M Tris–HCl, pH 8.8 (see Note 1). Store at room temperature.
2. Phenylmethanesulfonyl fluoride (PMSF) solution (100×): Prepare 0.1 M solution in ethanol and immediately freeze in 1.0-ml aliquots at −20°C.
3. Protease inhibitor cocktail (1,000×): 7 mM E-64, 1.5 mM bestatin, 2 mM pepstatin, 4 mM antipain, in DMSO. Store in aliquots −80°C.
4. Protein extraction buffer Y: 0.1 M Tris–HCl, pH 8.8, 2% (w/v) SDS, 5 mM EGTA, 10 mM EDTA. Store at room temperature.

Add 2% (v/v) β-mercaptoethanol (see Note 2), 1 mM PMSF, protease inhibitor cocktail right before use.

5. Tris-buffered phenol: Tris–HCl buffer-saturated phenol, pH 7.5–7.9. Store at 4°C.
6. 0.1 M Ammonium acetate in methanol: Store at −20°C in an explosion-proof freezer.
7. DIGE sample buffer: 7 M urea (see Note 3), 2 M thiourea, 4% (w/v) CHAPS. Filter through a 0.22-μm filter. Store in aliquots at −20°C.
8. Bio-Rad (Hercules) protein assay dye reagent concentrate (5×): Working solutions are prepared by diluting one part of dye concentrate with four parts of H_2O. Store at 4°C.
9. 50-ml Centrifuge tubes: VWR (Radnor, PA) superclear ultra-high performance centrifuge tubes with flat caps, polypropylene, maximum RCF 15,000 × *g*.

2.2. Preparation of IMAC Columns

1. Chelating Sepharose Fast Flow resin (GE Healthcare Bio-Sciences). Store at 4°C.
2. Free-flow chromatography columns, such as Bio-Rad Poly-Prep Chromatography Columns, reservoir 10 ml (Bio-Rad). Store at room temperature.
3. 0.1 M $GaCl_3$: Carefully break the glass seal of $GaCl_3$ powder (Sigma) and dissolve in deionized water (284 ml) in a glass beaker. Store at 4°C.
4. IMAC wash buffer: 6 M urea, 50 mM sodium acetate, pH 4.0, 0.25% (w/v) CHAPS. Pass through a 0.22-μm filter. Store at −20°C.

2.3. IMAC Chromatography

1. IMAC wash buffer: 6 M urea, 50 mM sodium acetate, pH 4.0, 0.25% (w/v) CHAPS. Pass through a 0.22-μm filter. Store at −20°C.
2. IMAC dilution buffer: 6 M urea, 50 mM sodium acetate, pH 4.0. Store at −20°C.
3. IMAC elution buffer: 6 M urea, 50 mM Tris–acetate, pH 7.4, 0.1 M EDTA, 0.1 M EGTA, 0.25% (w/v) CHAPS. Pass through a 0.22-μm filter. Store at −20°C.
4. 0.1 M ammonium acetate in methanol: Store at −20°C in an explosion-proof freezer.
5. DIGE sample buffer: 7 M urea (see Note 3), 2 M thiourea, 4% (w/v) CHAPS. Filter through a 0.22-μm filter. Store in aliquots at −20°C.
6. 15-ml Corex glass centrifuge tubes used with adaptors in a horizontal rotor.

7. Bio-Rad (Hercules, CA) protein assay dye reagent concentrate (5×). Store at 4°C.
8. 20% Ethanol.

2.4. Two-Dimensional Difference Gel Electrophoresis

1. CyDye DIGE fluor, Cy2-, Cy3-, and Cy5-minimal dyes (GE Healthcare Bio-Sciences).
2. Anhydrous dimethylformamide (DMF), Aldrich (22,705-6). Store at room temperature (see Note 4).
3. 1.5 M Tris–HCl, pH 8.8. Store at 4°C.
4. pH indicator strips (Sigma, P4536), pH test strips (pH 4.5–10.0).
5. 10 mM lysine; store in aliquots at −20°C.
6. DIGE sample buffer: 7 M urea (see Note 3), 2 M thiourea, 4% (w/v) CHAPS. Filter through a 0.22-μm filter. Store in aliquots at −20°C.
7. 1 M Dithiothreitol (DTT): Prepare fresh solutions for each use.
8. IPG buffer, pH 4–7 (GE Healthcare Bio-Sciences).
9. Immobilized pH gradient (IPG) strips (GE Healthcare Bio-Sciences), pH 4–7, 24 cm. Store at −80°C.
10. DryStrip cover fluid (GE Healthcare Bio-Sciences).
11. 30.8% (w/v) Acrylamide/bisacrylamide solution: Dissolve 900 g high-quality acrylamide powder (>99.9%, Bio-Rad) (see Note 5) and 24 g Bis (*N*,*N*′ methylenebisacrylamide) (Bio-Rad) in 1,500 ml H_2O. Mix by stirring, and add H_2O to total volume of 3,000 ml. Filter the solution through a 0.22-μm membrane to get rid of the insoluble substance. Store at 4°C in dark bottle and away from light.
12. 10% (w/v) SDS in H_2O.
13. Tris–HCl buffer: 1.5 M Tris–HCl, pH 8.8. Store at room temperature.
14. PlusOne ammonium persulfate (GE Healthcare Bio-Sciences).
15. PlusOne *N*,*N*,*N*,*N*-tetramethylethylenediamine (TEMED) (GE Healthcare Bio-Sciences). Store at 4°C.
16. Solution for casting 10% SDS-PAGE gels by Dalt six gel caster (total 380 ml): 95 ml of 1.5 M Tris–HCl, pH 8.8, 127 ml of 30.8% acrylamide/Bis solution, 3.8 ml of 10% SDS, 154 ml H_2O, 0.38 g ammonium persulfate, 118 μl TEMED. Add ammonium persulfate and TEMED right when ready to pour the gel.
17. Water-saturated isobutanol: Mix 800 ml of *n*-butanol with 200 ml of water in a glass bottle, shake violently, and settle to allow phase separation. Use the top layer to overlay gels. Store at room temperature.

18. SDS equilibration buffer: 6 M urea, 30% (v/v) glycerol, 2% (w/v) SDS, 50 mM Tris–HCl, pH 8.0; aliquot in 50-ml tubes and store at −20°C.
19. Agarose overlay solution: 0.5% (w/v) low melting point agarose (Sigma) in SDS electrophoresis running buffer, microwave to melt all the agarose, and keep it at 65°C water bath before use. Store at room temperature.
20. Protein molecular weight markers: Load a precut 5 × 10-mm filter paper strip with 3 μl of prestained protein marker (PageRule Prestained Protein Ladder, MW from 10 to 170 kDa, Fermentas, Glen Burnie, MA). Also add 1 μl of 0.05% Bromophenol blue. Store at −20°C.
21. Bromophenol blue (Sigma): 0.05% (w/v).
22. 1× SDS electrophoresis running buffer: 25 mM Tris, 192 mM glycine, 0.1% (w/v) SDS. Do not adjust pH. Store at room temperature.
23. 1× SDS electrophoresis running buffer: 25 mM Tris, 192 mM glycine, 0.1% SDS. Add 12.1 g of Tris base, 57.6 g of glycine, 4 g of SDS, and water to 4L. Store at room temperature.
24. SDS equilibration buffer: 6 M urea, 30% (v/v) glycerol, 2% (w/v) SDS, 50 mM Tris–HCl, pH 8.0; aliquot in 50-ml tubes and store at −20°C. Add DTT (final concentration: 0.5% w/v) or iodoacetamide (final concentration: 2% w/v) right before use.
25. 10% (w/v) SDS in H_2O.

2.5. Western Blot

1. 10× PBS solution: 1.37 M NaCl, 27 mM KCl, 100 mM Na_2HPO_4, 18 mM KH_2PO_4; adjust pH to 7.4 with HCl if necessary. Autoclave before storing at room temperature. Working solutions are prepared by dilution in H_2O.
2. PBST solution: Add Tween-20 to 1× PBS solution to final concentration of 0.1%.
3. Nitrocellulose membrane, 0.45 μm pore size (Bio-Rad).
4. Blocking solution: PBST solution with 5% (w/v) nonfat dry milk.
5. Antibody buffer: Blocking solution with diluted antibody. The working antibody concentration must be determined empirically.
6. Primary antibody (e.g., Rabbit anti-GFP polyclonal antiserum for experiment in this chapter).
7. Secondary antibody (Goat Anti-Rabbit Antibody Conjugated to Horseradish Peroxidase, Bio-Rad). Store at 4°C.
8. Semi transfer buffer: 25 mM Tris, 192 mM glycine, 20% (v/v) methanol. Do not adjust pH. Store at room temperature.

9. Chemiluminescence reagent, such as SuperSignal West Dura Extended Duration Substrate (Thermo Fisher Scientific). Store at room temperature or 4°C.

2.6. IP-DIGE

1. Protein immunoprecipitation (IP) buffer: 50 mM Tris–Cl, pH 7.5, 50 mM NaCl, 0.3 M sucrose, 0.5% (v/v) Triton X-100, 0.2 mM PMSF, 1× protease inhibitor (equivalent of Protease Inhibitor Cocktail for plant cell extracts, Product Number P 9599, Sigma; add before use).
2. Protein A agarose beads. Store at 4°C.
3. Antibody against the protein of interest or its fused epitope tag. Store at 4°C.
4. Miracloth (Calbiochem).
5. All materials described in Subheading 2.4.

3. Methods

3.1. Preparation of Total Protein Samples

1. Plant tissues: Dark-grown 4-day-old *Arabidopsis thaliana* plants expressing the mutant *bzr1-1D* gene fused to cyan fluorescence protein (mBZR1-CFP) (18) in Columbia-0 ecotype background. Harvest whole seedlings of both control- and brassinolide-treated samples. Freeze the tissue in liquid nitrogen immediately after harvest.
2. Grind the frozen tissue in liquid nitrogen with a porcelain mortar and pestle to a fine powder, weigh 2 g of powder from each sample, and transfer to a new 50-ml VWR tube for protein extraction and transfer the rest powder to another 50-ml VWR centrifuge tube for storing at −80°C freezer.
3. Keep the 50-ml tubes with 2 g of tissue powders on ice, and then immediately add 6.0 ml of extraction buffer Y. Do not let tissue thaw before adding the extraction buffer. Cover the tube and vortex for 1 min to mix the extraction solution with the tissue.
4. Incubate the 50-ml sample tubes in a tube holder in a water bath at 65°C for 15 min. Heating would increase protein solubilization by SDS.
5. Centrifuge at 14,000 × *g* for 20 min at 20°C to pellet the cell debris in an Eppendorf F-34-6-38 Fixed-Angle Rotor on an Eppendorf 5810R centrifuge.
6. Transfer 6.0 ml of supernatant to new 50-ml VWR tubes, add an equal volume of cold Tris–HCl-buffered phenol, and mix by inverting. Spin at 14,000 × *g* for 20 min at 4°C to separate the

mixture into three phases: water phase (top), phenol phase (bottom), and interphase (protein pellet).

7. Carefully remove the upper phase by pipetting (using 1.0-ml tips and avoiding disturbing the interphase). Keep the interphase and the bottom phase. Cover the 50-ml tubes tightly to avoid leaking. And discard the phenol waste in the waste container in compliance with the local environmental protection authority.
8. Add 6 ml of Tris–HCl buffer (50 mM Tris–HCl, pH 8.8) into lower phase, and mix by inverting. Spin at 14,000 × *g* for 20 min at 4°C. And discard the upper phase as in step 7.
9. Repeat step 8 one time.
10. To the lower phenol phase, add five volumes (~25 ml) of cold 0.1 M ammonium acetate in methanol. Mix and then incubate at –20°C overnight to precipitate proteins.
11. Centrifuge at 12,000 × *g* for 20 min at 4°C to pellet the protein. Discard the supernatant and do not disturb the pellet.
12. Gently add 5 ml of cold methanol. Leave on ice for 10 min. Swirl gently. Place the tubes back into the centrifuge rotor in the same orientation as before (to avoid disturbing the pellet), spin at 12,000 × *g* for 5 min, and remove all liquid with a pipette.
13. Repeat step 12 three more times. Remove trace supernatant, but do not dry the pellet (see Note 6).
14. Add 1.5 ml of DIGE sample buffer. Place the tubes back into the centrifuge rotor in the opposite orientation as before, and spin at 14,000 × *g* for 20 min to facilitate solubilizing the pellet.
15. Sonicate at the lowest power, 10% pulse, for 20 s, three times to increase protein solubilization. Keep on ice for 2 h to help solubilize protein. Spin at 14,000 × *g* for 5 min.
16. Transfer the supernatant to a 2-ml centrifuge tube, and then spin at 20,000 × *g* for 20 min at 4°C to remove pellet.
17. Determine protein concentration with Bio-Rad protein assay dye (see Note 7). Adjust protein concentration to equal (5–10 mg/ml) for each of the pair of samples. Aliquot the supernatant to 1.5-ml centrifuge tubes. The proteins can now be labeled with CyDye for 2-D DIGE, subject to IMAC fractionation, or stored at −80°C.

3.2. Preparation of IMAC Columns

1. Take 1 ml of Chelating Sepharose Fast Flow resin into each 10 ml of Chromatography Column. Wash with 10 ml of water. Repeat wash three times.
2. Add 9 ml of 100 mM $GaCl_3$, and incubate at room temperature overnight on a rotator to keep beads suspended. Columns can be stored at 4°C until needed.

3. Drain liquid and wash column four times with 10 ml of water each time. Then, wash with 5 ml of wash buffer.
4. Close the bottom of column with the yellow cap. Add 1 ml of wash buffer. Stir and pipette up–down to suspend beads. Dispense the beads into another empty column to make a pair of columns with the same amount of settled resin. Drain liquid when ready to load samples (see Note 8).

3.3. IMAC Chromatography

1. Mix 0.5 ml of protein (~5 mg) with 8 ml of dilution buffer in a 50-ml VWR centrifuge tube. Then, filter through a 0.22-μm filter by syringe.
2. Load the filtered samples (8.5 ml) onto the charged beads (0.5 ml of beads in each column), seal the column with the caps, and shake for 2 h on a roller shaker at room temperature (22°C).
3. Keep the column vertical and let the resin settle, then open the bottom, and let it flow through. Collect 0.5 ml of flow as a control.
4. Carefully add 2 ml of wash buffer onto the sides of the column, and let it flow through. Repeat this step four more times.
5. Carefully add 0.25 ml of elution buffer and let it flow through.
6. Close the bottom of column and add 0.25 ml of elution buffer. Incubate for 5 min, and then collect the elute.
7. Repeat step 6 four more times.
8. Combine the collected elute (total 1.25 ml) into a 15-ml Corex centrifuge tube. Precipitate the protein by adding 5× volume (6.25 ml) of cold 0.1 M ammonium acetate in methanol. Store at −20°C overnight.
9. Centrifuge at 15,000 × g (with adaptors) for 20 min at 4°C. Wash pellet with cold methanol. Resuspend protein pellet in DIGE buffer at ~5 mg/ml. Determine protein concentration with Bio-Rad Protein Assay buffer. The proteins are ready for labeling with CyDye or stored in −80°C freezer.
10. Regenerate column: Strip the beads of any bound metal by incubation with ten volumes of elution buffer for 2 h with gentle agitation. Drain, and wash with >20 volume water. Keep in 20% ethanol at 4°C.

3.4. Two-Dimensional Difference Gel Electrophoresis Analysis

1. Reconstitution of CyDye DIGE fluor minimal dyes in DMF: Dissolve one tube (5 nmol) of each dye (Cy3, Cy5; Cy2 is optional) in 25 μl of DMF. When usage is low and CyDyes need to be stored for a long time, we take out 1-nmol aliquots (5 μl) in new Eppendorf tubes, vacuum dry to remove all solvent, and store dry CyDyes at −80°C. Before use, we add 5 μl of fresh DMF to one dry aliquot of CyDye to reconstitute working solution. All dyes must be stored at −80°C freezer.

The dyes have a deep color: Cy2—yellow, Cy3—red, and Cy5—blue (see Note 9).

2. Labeling: Take 50 μg of protein (~6 μl) of each sample (see Note 10) into a new tube, and adjust the pH to 8–9 using about 0.3 μl of Tris–HCl buffer (1.5 M, pH 8.8). Mix each sample with 0.5 μl of Cy3 or Cy5 dye (see Note 11). Optional: Mix equal amount of all protein samples and then label the mixed samples with Cy2 to be used as internal reference.
3. Keep the reaction tube on ice in the dark for 2 h. Stop the reaction by adding 0.5 μl of lysine (10 mM) and incubate on ice for 10 min.
4. Follow the manufacturer's instruction to run the first-dimension IEF analysis. In this protocol, we use IPG strips (pH 4–7, 24 cm), strip holder (24 cm), and IPGPhor II, all from GE Healthcare. Prepare the rehydration mixture for loading: both mock and BL-treated protein samples (labeled with dyes), 9 μl of DTT (1 M), and 2.25 μl of IPG buffer (pH 4–7), adding DIGE buffer to 450 μl.
5. Gently pipette the rehydration solution evenly to the entire strip holder without introducing bubbles, and after carefully removing the plastic protective cover of the IPG strips, position the strip in the strip holder with gel side down (facing the rehydration solution), placing the acidic end of IPG strip over the positive electrode. Carefully overlay DryStrip Cover Fluid along the entire length of IPG strip (about 2 ml) to prevent evaporation and urea crystallization during electrophoresis.
6. Run the IEF as follows: rehydration for 2 h, 50 V for 10 h, step and hold at 500 V, 1,000 V for 1 h each, gradient to 8,000 V in 3 h, and then hold at 8,000 V, until reaching a total of 68, 000 V h. During the IEF, set the maximum current 50 μA/strip (see Note 12). IEF-focused strips can be stored at −80°C, but are preferably used immediately for second-dimension SDS-PAGE.
7. During IEF, cast large SDS-PAGE gels with Daltsix gel caster according to manufacturer's (GE Healthcare) instructions. Assemble the gel caster according to Ettan DALTsix user's manual (GE Healthcare). Combine the appropriate amount of acrylamide/Bis stock solution, 1.5 M Tris–HCl, pH 8.8, 2% SDS, and H_2O (see Note 13). Add appropriate amount of ammonium persulfate and TEMED, mix thoroughly by stirring, and pour the acrylamide gel solution into the gel caster until the gel surface is 1 cm below the upper edge of the small glass plate. Add water-saturated isobutanol to cover the acrylamide surface and let the solution polymerize at room temperature for at least 2 h. Best polymerization can be achieved by letting the gel polymerize overnight at room temperature.

Precast gels that are compatible with 2-D DIGE and Ettan Dalt II system can be obtained commercially (e.g., Jule Inc).

8. Equilibrate the focused IPG strips in equilibration buffer first with 0.5% DTT and then with 2% iodoacetamide, each for 15 min (see Note 14). Briefly rinse the equilibrated strips in SDS-PAGE running buffer, and then carefully transfer the strips on top of the large SDS-PAGE gels.
9. Heat the agarose overlay solution in a microware, and carefully pipette it onto the IEF strips. Use a long and even-edged thin spatula to push the IEF strips to rest on the gel, and make sure that no bubbles are introduced between the strip and the SDS-PAGE gel. Add a paper strip socked with molecular mass makers next to one side of the IPG strip.
10. Fill the lower part of Dalt II tank with 1× SDS electrophoresis buffer (~7 L) and cool it to 15°C. Then, slide the gel cassettes, one by one, into the tank. Fill open slots with the blank cassettes. Check that all cassettes are settled and then fill the upper part of the tank with 1× SDS electrophoresis buffer to the level below the maximum level set by the manufacturer (~2.5 L). Set the temperature to 15°C, and run the electrophoresis at 40 V for 2 h and then at 120 V until the bromophenol blue front just runs out of the gels. Stop the Ettan Dalt II electrophoresis system, and carefully remove the 2-DE gels from the tank.
11. Acquire images using a Typhoon scanner Trio or 9400 (GE Healthcare). Scan Cy2 images with the 488-nm laser and a 520/40-nm band-pass emission filter, Cy3 images using the 532-nm laser and 580/30-nm band-pass emission filter, and Cy5 images with the 633-nm laser and 670/30-nm band-pass emission filter. Set the focal plane +3 mm for Ettan DIGE gels. Set the appropriate PMT voltage and pixel size (see Note 15), and start the scan. Scanning two large Ettan Dalt II gels at 100 μm will take about 1 h.
12. Evaluation of phosphoprotein enrichment efficiency can be performed by comparing Western blot signals of input (total protein) with IMAC-enriched phosphoprotein when blotted to anti-phosphothreonine antibodies (Fig. 1a). Approximately tenfold increase in the phosphoprotein signal intensity was usually observed after IMAC enrichment in our studies. A representative 2-D DIGE image comparing total protein with phosphoprotein was shown in Fig. 1b, which showed dramatic differences in large numbers of protein spots.

3.5. Analysis of the DIGE Images with Decyder 6.5 Software

2-D DIGE images can be visualized in ImageQuant software (GE Healthcare). Differential spots can be visually identified as green or red color representing Cy3- and Cy5-labeled samples. Difference in phosphorylation of a protein between the two samples

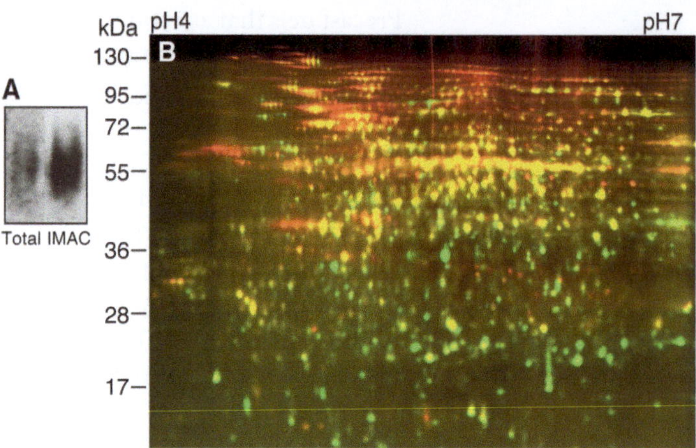

Fig. 1. Evaluation of IMAC enrichment of phosphoproteins. (**a**) Immunoblot of total protein and IMAC fraction using anti-phosphothreonine antibodies. (**b**) 2-D DIGE comparison of total protein (Cy3, *green*) to IMAC fraction (Cy5, *red*). Proteins that were enriched in the phosphoprotein fraction appear *red*. This research was originally published in ref. 15 and is reproduced with permission from the American Society for Biochemistry and Molecular Biology.

can be detected as a row of spots of similar molecular weight but different color from the acidic to basic side of the gel (see fig. 7 of ref. 15, and fig. 1 of ref. 19 for examples of images). To quantify the differences of spot intensity between the samples, images can be analyzed using the Decyder software (GE Healthcare).

1. Save the gel image files as Decyder DIGE file naming format and crop the images. The Decyder 6.5-recommended file name includes four parts: a general description (this part should be identical for three images from the same gel), followed by a functional description (standard, control, or treated), then labeling description (Cy2, Cy3, or Cy5), and an extension in the end. For example, three images from the same gel could be named as BLtreatment1 standard Cy2.gel, BLtreatment1 control Cy3.gel, and BLtreatment1 treated Cy5.gel. These three images are grouped into a folder named BLtreatment1. Use ImageQuant 5.3 (GE Healthcare Bio-Sciences) or ImageQuant TL software to crop image files to exclude nonessential information. We usually only keep the areas with well-resolved spots and remove the four boundaries.
2. Analyze each gel images using the Decyder DIA (differential in-gel analysis) module for spot detection and quantitation for intra-gel analysis. For a well-resolved Ettan Dalt II large gel, set the estimated number of spots as 5,000. And exclude those spots with slope > 1.1, area < 100, volume < 1,000, or peak < 100. Also only include those areas with clearly visible spots.

3. Process multiple DIGE gels with biological variation analysis (BVA) module to perform gel-to-gel matching of spots, and to quantify protein expression across multiple gels. Group the gel images into three groups: standard, control, and treated samples. After spot matching, statistical analysis, such as Student's *t*-test and average ratio, should be performed between the control and treated groups. Set the average ratio (>1.3 or <1.3, in our study) and Student's *t*-test value ($p < 0.05$) to identify spots of interest.
4. Manually check spots of interest to make sure that they are correctly matched among different gels, and they are real protein spots and not dust particles (see Note 16). Spots of interest can be picked by robotic Ettan Spot Picker from preparative gels (see Note 17), in-gel trypsin digested, and analyzed by standard tandem mass spectrometer for spot identity.

3.6. Detection of Multiple Phosphorylation Forms by 2-DE Western Blotting

The extensive spot overlap in 2-DE gels demands that the identity of important protein spots be confirmed by immuno methods. 2-DE Western blotting can analyze modification forms of a given protein (see fig. 1c of ref. 19, e.g.). Immunoblot of 2-D DIGE gel allows alignment of immunosignals to the DIGE spot patterns.

1. Carefully open the 2-DE gel cassette with a Wonder Wedge (GE Healthcare), and allow the gel to stick to one glass plate. Carefully cut the gel area that contains the expected protein with a plastic ruler.
2. Electro-blot the SDS-PAGE gel to nitrocellulose membrane.
3. Incubate the nitrocellulose sheet in the immunoblocking solution at 4°C overnight with gentle shaking.
4. Transfer membrane to first antibody solution (1:500 dilution in our case); incubate for 1 h at room temperature with gentle shaking.
5. Wash the blot four times, 5 min each, in PBST with shaking.
6. Transfer blot to appropriately diluted secondary antibody solution (1:500 dilution in our case). Incubate with shaking for 60 min.
7. Wash the blot four times, 5 min each, in PBST with shaking.
8. Prepare chemiluminescence (e.g., SuperSignal West Dura) working solution by mixing equal volumes of the Stable Peroxide Solution and the Luminol/Enhancer Solution. Use appropriate amount of working solution to cover the entire membrane.
9. Remove blot from Working Solution and place it in a plastic membrane protector or plastic wrap. Remove excess liquid using an absorbent tissue.

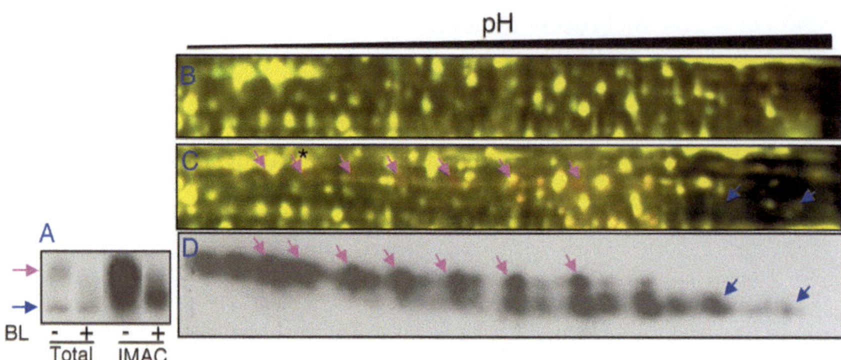

Fig. 2. BZR1 protein is detected by 2-D DIGE following IMAC enrichment of phosphoproteins. (**a**) Immunoblot of mBZR1-CFP in total protein and IMAC fractions of plants treated with mock solution (–) or 100 nM BL (+) for 2 h using anti-GFP antibodies. (**b**, **c**) Total protein (**b**) or phosphoprotein fractions (**c**) were analyzed by 2-D DIGE after labeling BL-treated and untreated samples with Cy3 and Cy5, respectively. (**d**) Gel of *panel* **c** was blotted and probed with anti-GFP antibodies. *Cyan arrows* show the hyperphosphorylated and *blue arrows* show the hypophosphorylated or unphosphorylated mBZR1-CFP proteins. Matching areas of 2-DE are shown in *panels* **b–d**. This research was originally published in ref. 15 and is reproduced with permission from the American Society for Biochemistry and Molecular Biology.

10. Acquire both luminescence image and Cy dye image from the blot using an imaging system equipped with sensitive CCD camera (such as the FluroChem Q imaging system of Cell Biosystems, formerly Alpha-Innotech). Image can also be acquired using film, and then aligned to the blot to locate the spots on gel.
11. Enrichment of BZR1-CFP was shown by immunoblotting to an anti-GFP antibody (Fig. 2a). In our study, dephosphorylation of BZR1-CFP after brassinolide treatment was not detected in 2-D DIGE of total proteins (Fig. 2b), but was detected as a string of red and green spots shifted along IEF dimension in the phosphoprotein samples (Fig. 2c). Immunoblotting of 2-DE gel detected BZR1 in the corresponding spots (Fig. 2d).

3.7. Verifying Changes of Protein Modification/Phosphorylation Using IP-DIGE

Changes in posttranslational modifications that shift the spot position, such as phosphorylation, can be verified by immunoprecipitation followed by 2-D DIGE analysis. A protein of interest is immunoprecipitated using antibody against the protein itself or the epitope tag fused to the protein expressed in transgenic plants. The protein immunoprecipitated from different biological samples can be labeled with Cy2, Cy3, and Cy5, and analyzed by 2-D DIGE. Difference in phosphorylation, or other modifications that change the charge of the protein, can be detected as spot shift along the IEF dimension and quantified by spot ratios along the train of spots. A control sample lacking the antigen protein, such as a knockout mutant or nontransgenic plant, can be labeled with Cy2 and included to confirm the identity of the protein spots. IP-DIGE directly compares biological samples for the amount of different modification forms of a protein in the same 2-DE gel, which is more reliable than comparing two gels in 2-DE Western blotting described above.

3.7.1. Prepare Antibody-Protein A Beads

1. Wash 50 μl of 1:1 protein A beads slurry with 1 ml of IP buffer. Keep about 100 μl of buffer and add 10 μg of antibody against the protein of interest or its epitope tag. Incubate at 4°C overnight or longer with gentle shaking.
2. Before the protein extract is ready (step 6 in Subheading 3.7.2 below), centrifuge at 1,000 × *g* for 1 min to collect the antibody-Protein A beads. Add 1 ml of IP buffer and split equally into two 1.5-ml tubes. Centrifuge again and remove all liquid.

3.7.2. Extraction of Proteins

1. Grind tissues of treated and untreated samples into fine powder in liquid nitrogen.
2. Weigh 0.5 g of tissue powder in tubes and add 2 ml of IP buffer.
3. Vortex for 1 min and keep on ice for 5 min.
4. Filter the extracts through two layers of Miracloth into 50-ml tubes, and centrifuge at 3,700 × *g* for 3 min.
5. Transfer the supernatant into new 2-ml tubes.
6. Centrifuge at 10,000 × *g* for 15 min at 4°C.
7. Transfer supernatant into new 2-ml tubes.

3.7.3. Immunoprecipitation

1. Add 1.5 ml of extract of control and treated samples to each tube of antibody-Protein A beads.

Incubate with shaking at 4°C for 90 min.

2. Spin at 1,000 × *g* for 30 s, and remove supernatant.
3. Wash four times using 1 ml of IP buffer containing 0.1% Triton X-100 (see Note 18).
4. Wash with 1 ml of 50-mM Tris–Cl, pH 8.5. Spin at 1,000 × *g* for 30 s. Remove all liquid.
5. Add 10 μl of DIGE buffer (pH 8.5) to the beads (see Note 19).

3.7.4. CyDye Labeling and 2-D DIGE

1. Add 50 pmol of Cy3 and Cy5 minimum dye (see Note 20) to the IP beads of control and treated samples, respectively. Mix by tapping the tube. Shake tubes at 4°C in the dark overnight. Sensitivity of dye labeling can be improved with saturation CyDye labeling (see Note 20).
2. Add 10 μl of 10 mM lysine to stop the labeling. Incubate on ice for 10 min.
3. Add 50 μl of DIGE buffer. Spin at top speed in Eppendorf centrifuge for 10 s. Transfer liquid to new 0.5-ml Eppendorf tubes.

4. Repeat step 3 and combine the eluates.
5. Combine samples labeled with different CyDyes. Add DIGE buffer to bring to the final volume suggested for the IPG strip to be used (e.g., 250 μl for 7-cm IPG strip).
6. Run 2-D DIGE as described above. Choose IPG pH range based on the predicted pI of the protein or observed spot location in initial 2-D DIGE, where the protein was identified (if nontagged native protein is immunoprecipitated by its own antibody). Phosphorylated form of the protein is more acidic.
7. Scan gels for each CyDye image. Adjust the contrast of each channel so that the spots of IgG are yellow. The different phosphorylated forms of the protein should form a row of spots, and green or red color at either end of the row would indicate difference in phosphorylation or other charge-changing modification between the control and treated samples. The relative level of modification forms can be quantified using the DeCyder software, as shown by Deng et al. (17).

4. Notes

1. Use high-quality water for all steps, such as double-distilled water or commercial LC/MS grade bottled water (Sigma).
2. Be sure to add the reducing agent β-mercaptoethanol; otherwise, the purified total protein will be contaminated with chlorophyll and other compounds.
3. Prepare urea-containing solution freshly and do not heat it over 37°C since high temperature converts urea to isocyanate which modifies proteins by carbamylation of primary amines. This modification can lead to changes in protein pI and result in spot train (a number of spots on the horizontal line) on 2-DE gels.
4. DMF must be anhydrous since it reacts with water and degrades to produce amine compounds over time after opening. Amines react with the NHS ester of CyDye, reducing the amount of dyes available for protein labeling. We usually take DMF out from an unopened bottle with syringe and needle without removing the rubber cap. Aliquot DMF into 1.5-ml centrifuge tubes for single-time use. Aliquots of DMF can be kept at −80°C freezer for at least 1 year.
5. Use high-purity acrylamide for 2-DIGE analysis, such as 99.9% pure acrylamide powder (Bio-Rad), ≥99% (HPLC) acrylamide (Sigma), or PlusOne Acrylamide PAGE (GE Healthcare Bio-sciences). Low-quality acrylamide contains impurities that interfere with gel polymerization reaction and produce high fluorescence backgrounds.

6. Do not dry the protein pellet; otherwise, it will be very difficult to dissolve it. Remove the final few drops of liquid in the tube with fine tips, and then add DIGE buffer immediately to cover the pellet.
7. High concentration of DIGE buffer interferes with protein concentration determination by Bio-Rad protein assay. In our case, 1 μl of DIGE buffer in 1,000 μl of Bio-Rad protein assay buffer (working solution) is appropriate for quantification.
8. Pay attention to details to make sure that the two columns are identical. The IMAC columns should have the same amount of beads, packed in the same way, and the same flow rate.
9. The CyDye in DMF gives a deep color: Cy2—yellow, Cy3—red, and Cy5—blue. Notice that the color of Cy3 and Cy5 solutions is opposite to the color of their fluorescence.
10. It is recommended by manufacturer (GE Healthcare) to use 50 μg of protein and 400 pmol of dye in each reaction. However, we found that less dye (50–100 pmol) or less protein (20–30 μg) yield sufficient signal and can significantly reduce the cost for 2-D DIGE experiment. But the gel plates must be absolutely clean to reduce background signal.
11. Since fluorescent properties of DIGE dyes are affected by exposure to light, it is recommended that protein labelling should be performed under dim light, labeling reactions should be kept in the dark on ice, and exposure of labeled protein to light should be kept to minimum.
12. The length of the IPG strip and its pH range should be determined according to the sample of interest. In general, using both the 24-cm IPG strips (pH 4–7) and 18-cm IPG strips (pH 6–11) gives a good balance of coverage and resolution.
13. Make sure that all solutions for preparing large 2-DE gels are at room temperature before use. Otherwise, the gels are not polymerized well, and proteins (especially of those with high molecular mass) are not resolved well on the 2-DE gel. Since acrylamide stock and Tris buffer solution are usually stored at 4°C, it is a good practice to take out the required amount of Tris–HCl, acrylamide, and water, mix them, and leave at room temperature overnight before use.
14. Make sure that the equilibration buffer reaches room temperature before use. Otherwise, equilibration will not be achieved adequately. Add DTT and iodoacetamide powder in the equilibration buffer right before use. DTT is easily dissolved in the SDS equilibration buffer, but iodoacetamide does not dissolve immediately. So iodoacetamide should be added into the SDS equilibration buffer at least 20 min before use.

15. The PMT voltage can be set from 300 to 1,000 V, but we recommend using voltage below 600 V to reduce fluorescence background. A quick prescan at 500- or 1,000-μm pixel resolution should be performed to identify a suitable voltage, and adjust the voltage as needed to have a maximum pixel intensity of around 50,000–90,000 on the gels. Scan the images at 100-μm pixel resolution for Decyder software analysis. Images scanned at higher resolution (50 or 25 μm) have larger file size, but are not compatible with Decyder 6.5.
16. If protein extraction, IEF, and second-dimension SDS-PAGE are performed in the same batch and the gels are prepared simultaneously in the same gel caster, Decyder will easily match spots among different gels. Gels run at different batches usually need laborious manual matching. So it is recommended to perform protein extraction and 2-DIGE in a batch for sample comparison.
17. For minimum dye-labeling 2-D DIGE, the spot signal level is determined by the amount of CyDye and not the amount of protein used, except that the protein concentration affects labeling efficiency. Thus, we use low amount of protein loading (50–100 μg/gel) to obtain highly resolved images in analytical gels. At least three biological repeats are required for statistical analysis. For spot picking and protein identification by mass spectrometry, we suggest maximum amount of protein loading, and we usually run 0.2–1 mg protein in a 24-cm preparative gel for spot picking. However, high amount of protein loading on the 2-DE gels usually increases spot overlap and lowers spot resolution. For some spots in crowded area, we run narrow pH range gels to improve separation. Because CyDye shifts spots to higher molecular weight, spot picking by robotic Ettan Spot Picker should be based on post-gel staining with Deep Purple or SYPRO ruby fluorescence dyes. Manual picking is possible after staining gel with silver or Coomassie blue, though spots must be carefully matched to the DIGE image.
18. At this step, an aliquot of the IP beads containing the putative phosphorylated protein can be treated with phosphatase (calf intestinal phosphatase, CIP), and then processed for labeling using Cy2 and DIGE analysis in the same gel as untreated samples. CIP treatment should dephosphorylate the protein and shift the phospho-spots to the basic side, confirming that spot shift is due to phosphorylation.
19. For efficient CyDye labeling, minimize the volume and maximize the concentration of protein. Make sure that pH is 8.5–8.8.
20. Using saturation-labeling CyDyes, DIGE Fluor can be over 100-fold more sensitive than minimal labeling CyDyes. When using saturation-labeling CyDyes, beads should be resuspended

in the buffer for saturation CyDyes. Only proteins containing cysteine can be detected by saturation-labeling CyDyes, and only proteins containing lysine are detected by minimal CyDyes.

Acknowledgments

This work was supported by the Division of Chemical Sciences, Geosciences, and Biosciences, Office of Basic Energy Sciences of the U.S. Department of Energy through Grant DE-FG02-08ER15973 (to Z.-Y. W. and A. L. B.). Z. D. was partially supported by funds from Zhejiang Academy of Agricultural Sciences.

References

1. Hunter T (2007) The age of crosstalk: phosphorylation, ubiquitination, and beyond. Mol Cell 28:730–738
2. Kersten B, Agrawal GK, Durek P, Neigenfind J, Schulze W, Walther D, Rakwal R (2009) Plant phosphoproteomics: an update. Proteomics 9:964–988
3. Ross PL, Huang YN, Marchese JN, Williamson B, Parker K, Hattan S, Khainovski N, Pillai S, Dey S, Daniels S, Purkayastha S, Juhasz P, Martin S, Bartlet-Jones M, He F, Jacobson A, Pappin DJ (2004) Multiplexed protein quantitation in *Saccharomyces cerevisiae* using amine-reactive isobaric tagging reagents. Mol Cell Proteomics 3:1154–1169
4. Ong SE, Blagoev B, Kratchmarova I, Kristensen DB, Steen H, Pandey A, Mann M (2002) Stable isotope labeling by amino acids in cell culture. SILAC, as a simple and accurate approach to expression proteomics. Mol Cell Proteomics 1:376–386
5. Gygi SP, Rist B, Gerber SA, Turecek F, Gelb MH, Aebersold R (1999) Quantitative analysis of complex protein mixtures using isotope-coded affinity tags. Nat Biotechnol 17:994–999
6. Macek B, Mann M, Olsen JV (2009) Global and site-specific quantitative phosphoproteomics: principles and applications. Annu Rev Pharmacol Toxicol 49:199–221
7. Nuhse TS, Bottrill AR, Jones AM, Peck SC (2007) Quantitative phosphoproteomic analysis of plasma membrane proteins reveals regulatory mechanisms of plant innate immune responses. Plant J 51:931–940
8. O'Farrell PH (1975) High resolution two-dimensional electrophoresis of proteins. J Biol Chem 250:4007–4021
9. Unlu M, Morgan ME, Minden JS (1997) Difference gel electrophoresis: a single gel method for detecting changes in protein extracts. Electrophoresis 18:2071–2077
10. Tonge R, Shaw J, Middleton B, Rowlinson R, Rayner S, Young J, Pognan F, Hawkins E, Currie I, Davison M (2001) Validation and development of fluorescence two-dimensional differential gel electrophoresis proteomics technology. Proteomics 1:377–396
11. Tan HT, Tan S, Lin Q, Lim TK, Hew CL, Chung MC (2008) Quantitative and temporal proteome analysis of butyrate-treated colorectal cancer cells. Mol Cell Proteomics 7:1174–1185
12. Thon JN, Schubert P, Duguay M, Serrano K, Lin S, Kast J, Devine DV (2008) Comprehensive proteomic analysis of protein changes during platelet storage requires complementary proteomic approaches. Transfusion 48:425–435
13. Wu WW, Wang G, Baek SJ, Shen RF (2006) Comparative study of three proteomic quantitative methods, DIGE, cICAT, and iTRAQ, using 2D gel- or LC-MALDI TOF/TOF. J Proteome Res 5:651–658
14. Thelen JJ, Peck SC (2007) Quantitative proteomics in plants: choices in abundance. Plant Cell 19:3339–3346
15. Tang W, Deng Z, Oses-Prieto JA, Suzuki N, Zhu S, Zhang X, Burlingame AL, Wang ZY (2008) Proteomics studies of brassinosteroid

signal transduction using prefractionation and two-dimensional DIGE. Mol Cell Proteomics 7:728–738

16. Gu Y, Deng Z, Paredez AR, DeBolt S, Wang ZY, Somerville C (2008) Prefoldin 6 is required for normal microtubule dynamics and organization in *Arabidopsis*. Proc Natl Acad Sci USA 105:18064–18069
17. Deng Z, Zhang X, Tang W, Oses-Prieto JA, Suzuki N, Gendron JM, Chen H, Guan S, Chalkley RJ, Peterman TK, Burlingame AL, Wang ZY (2007) A proteomics study of brassinosteroid response in *Arabidopsis*. Mol Cell Proteomics 6:2058–2071
18. He JX, Gendron JM, Yang Y, Li J, Wang ZY (2002) The GSK3-like kinase BIN2 phosphorylates and destabilizes BZR1, a positive regulator of the brassinosteroid signaling pathway in *Arabidopsis*. Proc Natl Acad Sci USA 99:10185–10190
19. Tang W, Kim T-W, Oses-Prieto JA, Sun Y, Deng Z, Zhu S, Wang R, Burlingame AL, Wang Z-Y (2008) BSKs mediate signal transduction from the receptor kinase BRI1 in *Arabidopsis*. Science 321:557–560

Chapter 5

Quantitative Analysis of Plasma Membrane Proteome Using Two-Dimensional Difference Gel Electrophoresis

Wenqiang Tang

Abstract

The plasma membrane (PM) controls cell's exchange of both material and information with the outside environment, and PM-associated proteins play key roles in cellular regulation. Numerous cell surface receptors allow cells to perceive and respond to various signals from neighbor cells, pathogens, or the environment; large numbers of transporter and channel proteins control material uptake or release. Quantitative proteomic analysis of PM-associated proteins can identify key proteins involved in signal transduction and cellular regulation. Here, we describe a protocol for quantitative proteomic analysis of PM proteins using two-dimensional difference gel electrophoresis. The protocol has been successfully employed to identify new components of the brassinosteroid signaling pathway, and should also be applicable to the studies of other plant signal transduction pathways and regulatory mechanisms.

Key words: Signal transduction, Proteomics, Plasma membrane, 2-D DIGE, Protein prefractionation, Brassinosteroids

1. Introduction

Signal transduction is important for plants because plants are sessile and rely on signal transduction pathways to regulate gene expression and to adapt to the changing environment. However, our knowledge of plant signal transduction pathways is still very limited. Although genetic studies in model systems such as *Arabidopsis* have identified key components of many signaling pathways, there are often gaps in these genetically defined pathways largely because many genes' functions cannot be identified by traditional genetic studies due to genetic redundancy or lethality. On the other hands, recent progresses in proteomic technologies have provided important tools for identifying new signal transduction proteins (1).

Zhi-Yong Wang and Zhenbiao Yang (eds.), *Plant Signalling Networks: Methods and Protocols*, Methods in Molecular Biology, vol. 876, DOI 10.1007/978-1-61779-809-2_5,

Compared with the traditional genetic studies which rely on phenotypes caused by altering gene activity, proteomic studies identify protein functions based on the changes of protein's abundance or modification associated with cellular or physiological perturbation or protein's interaction with other known proteins. Proteomics not only identifies proteins' function, but also reveals their mechanisms of function and regulation, such as posttranslational modifications and protein–protein interaction (2, 3).

While proteomics is a powerful approach for biological research, it has its own challenges and limitations. A major challenge is the high complexity of the proteomes. The most recent release of *Arabidopsis* genome annotation indicates that the *Arabidopsis* genome contains 27,416 protein-encoding genes. Each gene can encode different proteins due to alternative splicing and each protein can be regulated by a wide range of posttranslational modifications, such as phosphorylation, methylation, ubiquitination, and acetylation. Therefore, the complexity of cellular protein species is actually much higher. Moreover, the abundance of the proteins in a cell may differ by up to ten orders of magnitude, and proteins involved in signal transduction or regulation normally exist in low abundance and thus are difficult to detect. Due to high sample complexity and limited analytical power, current proteomic technologies cannot analyze all the proteins of a eukaryotic sample, unlike genomic analysis of the whole genome by microarray or high-throughput sequencing. Therefore, reducing the complexity of protein profiles by sample prefractionation strategies, such as subcellular fractionation and affinity purification, is considered critical for studying low-abundance signal transduction proteins by proteomic approaches (4, 5).

Two types of proteomic approaches have been widely used. Two-dimensional gel electrophoresis (2-DE) separates proteins based on charge by isoelectric focusing (IEF) electrophoresis in first dimension and based on size by SDS-PAGE in second dimension (6). Traditional 2-DE separates one sample in each gel and relative quantification is achieved by comparison between gels, which is not accurate because of gel-to-gel variations. A significant improvement was introduced about a decade ago by a new method called 2-D difference gel electrophoresis (2-D DIGE) (7, 8). 2-D DIGE uses different fluorescence dyes to covalently label proteins before proteins are mixed together and then separated in the same 2-DE gel. Images of different samples acquired from the same gel have the same spot patterns and can be directly compared to quantify abundance ratios. 2-D DIGE allows direct comparison of up to three samples within the same gel. By using a pooled reference sample in all gels, cross-gel comparison and multiplexing can be performed when large numbers of samples need to be compared (8).

The proteins of spots of interest can be identified by in-gel digestion followed by mass spectrometry analysis.

Another proteomic approach is gel-free and based on direct mass spectrometry analysis of digested peptides. This approach has become more and more popular due to improvement of MS technology, such as powerful tandem mass spectrometry (MS/MS). Typically, liquid chromatography of tryptic peptides is performed before MS analysis (LC–MS/MS), and isotope or isobaric labeling improves quantitative comparison between samples (3).

Although there has been a trend in the field to shift from 2-DE to gel-free MS approaches, each method has its own strength and limitations. 2-DE is believed to have limitations for certain types of proteins, such as hydrophobic proteins and proteins of extreme molecular weight or charge. In particular, 2-DE is widely considered ineffective for separating transmembrane proteins; however, proteins containing single transmembrane domain or peripheral membrane-associated protein can be readily resolved in 2-DE. Gel-free methods in principle can detect any proteins/peptides. However, in reality, peptides representing only a few thousands of proteins can be detected in typical LC–MS/MS analysis. Major advantages of 2-D DIGE include reliable quantitative analysis of not only protein abundance but also posttranslational modifications/processings that cause spot shifts in the gel. For example, phosphorylation shifts a protein spot to the acidic side in the IEF dimension and protein cleavage or modification by ubiquitination would shift protein spot along the SDS-PAGE dimension (9). In LC–MS/MS, a change of posttranslational modification can only be analyzed when the peptide containing the modification is detected, and usually only a few peptides are detected in LC–MS/MS for each low-abundance protein. Therefore, 2-D DIGE has some unique advantages in studying signal transduction.

Following the protocol described here, we have analyzed the early brassinosteroid (BR)-regulated plasma membrane (PM) proteins using 2-D DIGE, and we were able to identify two major components of the BR signal transduction pathway: the BAK1 receptor kinase previously identified by molecular genetics and the BR-signaling kinases (BSKs) as new components that transduce signal downstream of the BR receptor kinase BRI1. We also identified a number of other novel BR-regulated PM-associated proteins (5, 10). Our data demonstrated that 2-D DIGE coupled with prefractionation is a very powerful approach in studying signal transduction pathway. Although the protocol is described for comparison between BR-treated and untreated samples, it can be used for comparison between any biological samples.

2. Materials

2.1. Growth of Arabidopsis Seedlings and Cell Lysis

1. Growth media: 1/2 Murashige and Skoog basal salt mixture and 1.5% sucrose; adjust pH to 5.7 with 1 M KOH.
2. 2 mM brassinolide (BL) in 85% ethanol; aliquote and store at −20°C.
3. 1,000× Protease inhibitor cocktail: 1 mM E-64 (Sigma), 1 mM bestatin (Sigma), 1 mM pepstatin (Sigma), and 2 mM leupeptin (Sigma) in dimethyl sulfoxide (DMSO); aliquote and store at −80°C.
4. 1,000× Phenylmethanesulfonyl fluoride (PMSF, Sigma): Dissolve 1 M PMSF in isopropanol. Store in aliquots at −20°C.
5. 100 mM cantharidin (Sigma) in DMSO; aliquote and store at −20°C.
6. 1 M dithiothreitol (DTT, Sigma) in double-distilled water (ddH_2O); aliquot and store at −20°C.
7. 200 mM sodium orthovanadate in ddH_2O (see Note 1).
8. Tissue grinding buffer (buffer H): 25 mM HEPES, pH 7.5 with NaOH, 0.33 M sucrose, 10% glycerol, 0.6% polyvinylpyrrolidone, 5 mM ascorbic acid, 5 mM Na-EDTA, 50 mM NaF, 2 mM imidazole, and 1 mM sodium molybdate. Store aliquots at −20°C and thaw on ice before use (see Note 2). Right before grinding, add protease inhibitor cocktail, PMSF, DTT, cantharidin, and activated sodium orthovanadate stock solution to make the final grinding buffer containing 1 μM PMSF, 5 mM DTT, 0.1 mM cantharidin, and 1 mM sodium orthovanadate.

2.2. Plasma Membrane Isolation

1. Dextran stock solution (20%, w/w) is prepared by adding 780 g of ddH_2O to 220 g of Dextran T-500 (average molecular weight 500,000, Fisher scientific) on a balance. The mixture can be incubated at 50°C water bath to facilitate dissolving. Aliquot and store at −20°C (see Note 3).
2. PEG stock solution (40%, w/w) is prepared by adding 300 g of ddH_2O to 200 g of polyethylene glycol (PEG)-3350 (average molecular weight 3,350, Sigma). Store in aliquots at −20°C.
3. Potassium phosphate buffer (0.2 M, pH 7.8): Mix 0.2 M K_2HPO_4 with 0.2 M KH_2PO_4 by constant stirring until the pH of the solution reaches 7.8.
4. 1 M potassium chloride in ddH_2O.
5. Na-EDTA 50 mM in ddH_2O, pH 7.8 with NaOH.
6. 2 M Sucrose in ddH_2O.
7. Microsome resuspension solution (buffer R): Prepare 4× stock containing 20 mM potassium phosphate (pH 7.8), 1 M

Table 1
Two-phase system with 6.0% Dextran T-500, PEG 3350, and 8 mM KCl

	Phase mixture		
	4 g	**16 g**	**20 g**
[a]20% (w/w) Dextran T-500 (g)	1.2	4.8	6
[a]40% (w/w) PEG 3350 (g)	0.6	2.4	3
4× Buffer R (g)	0.75	3	3.75
ddH_2O to the weight of (g)	3	12	15
Microsome in buffer R (or buffer R alone) to the weight of (g)	4	16	20

[a]Two-phase systems with various Dextran concentrations can be prepared by varying the amount of 20% Dextran, 40% PEG 3350 and ddH_2O

sucrose, 32 mM KCl, and 0.4 mM EDTA using stock solution listed on items 3–6. For resuspension of microsome, add protease inhibitor cocktail and 1 mM DTT to 1× buffer R right before use. KCl concentration in buffer R should match the final KCl concentration chosen for the two-phase mixture. In this case, the author used a two-phase mixture with 6.0% polymers and 8 mM KCl.

8. T3 upper-phase dilution solution (buffer D): 5 mM potassium phosphate (pH 7.8) and 0.1 mM EDTA. Add protease cocktail and 1 mM DTT right before use.
9. Two-phase mixture preparation: One day before PM purification, according to Table 1, on a balance, mix 12 g of 20% (w/w) Dextran T-500, 40% (w/w) PEG 3350, 4× buffer R, and ddH_2O in a 25-ml glass centrifuge tube for a 16 g of two-phase mixture. Three 12 g of two-phase mixtures should be prepared for each sample and labeled with T1, T2, and T3, since two wash steps are to be used for PM purification. Leave T1 for future microsomal resuspension solution, and weight 4 g of 1× buffer R to each T2 and T3 tube to make a full strength 16 g of two-phase mixture. Vortex and spin T1, T2, and T3 at 1,000 × g for 5 min in a swinging rotor to facilitate phase settling (see Note 4). Store T1, T2, and T3 at 4°C until used for PM preparation.

2.3. Sample Labeling and Two-Dimensional Gel Electrophoresis

Ideally, two-dimensional electrophoresis should be performed using the 2-D DIGE system of GE Healthcare. 2-DE equipment of other companies, such as Bio-Rad, can also run 2-D DIGE with some modifications: the glass plates for SDS-PAGE should be low-fluorescence glass and one of the two glass plates for each gel must be treated with Bind Silane so that the gel will attach to one plate and will not change shape during staining and spot picking. Precast SDS-PAGE gels are commercially available, but we routinely cast our own gels to reduce cost. We recommend using the 24-cm IPG strips and 26 x 20-cm SDS-PAGE gels because large gels produce higher resolution. In general, the number of protein spots that can be detected is correlated to the size of the gels. A fluorescence scanner, such as the Typhoon Trio (GE Healthcare) or similar equipment, is required for acquiring the Cydye images. Image analysis should be performed using DeCyder software (GE Healthcare). Other 2-DE software can be used, but software that performs spot detection on separate images from the same gel tends to introduce errors in quantitation. Spot picking should be performed using the Ettan Spot picker of GE Healthcare, which is designed to pick spots based on fluorescence images with two reference spots. Manual spot picking is possible by post-staining the gel with Coomassie Blue or silver after acquiring the fluorescence images.

1. 2-DE buffer: 7 M urea, 2 M thiourea, and 4% CHAPS in ddH_2O. Filter through a 0.22-µm filter and store at −20°C as 1 ml of aliquots to avoid repeated freeze and thaw.
2. CyDye working solution: 0.1 mM CyDye DIGE Fluor minimal dyes (GE Healthcare) in high quality anhydrous dimethylformamide (DMF, Sigma) (see Note 5).
3. 10 mM Lysine in ddH_2O. Aliquot and store at −20°C.
4. SDS equilibration buffer: 6 M urea (BioUltra, for molecular biology, >99.5%, Sigma), 30% (v/v) glycerol, 2% (w/v) sodium dodecyl sulfate (SDS, >99%, Sigma), and 50 mM Tris–HCl, pH 8.8. Aliquot and store at −20°C. Add 0.5% (w/v) DTT or 4.5% (w/v) iodoacetamide (GE Healthcare) right before use.
5. Bind-Silane (γ-methacryloxypropyltrimethoxysilane) working solution: Dissolve 20 µl of Bind-Silane solution (GE Healthcare) in 80% (v/v) ethanol and 0.02% (v/v) glacial acetic acid. Prepare right before use.
6. Low-fluorescence glass plates for SDS-PAGE.
7. 1.5 M Tris–HCl, pH 8.8.
8. 10% (w/v) SDS in ddH_2O.
9. Acrylamide stock solution (30.8% w/v): Dissolve 300 g of high-quality acrylamide (for molecular biology, >99.5%, Sigma) and 8 g of *N*,*N*′-methylenebisacrylamide (Sigma) in 800 ml ddH_2O. Add ddH_2O to the volume of 1,000 ml and filter the solution through a 0.2-µm membrane to remove any insoluble substance. Store at 4°C and away from light.

10. 10% (w/v) Ammonium persulfate in ddH_2O. Aliquot and store at −20°C.
11. Water-saturated isobutanol: Mix 800 ml of isobutanol with 200 ml of water in a glass bottle, shake violently, and sit still to allow liquid separation. Use the top layer. Store at room temperature.
12. SDS Running buffer (10×): 250 mM Tris, 1,920 mM glycine, and 1% (w/v) SDS. Store at room temperature.
13. Agarose overlay solution: 0.5% (w/v) low melting agarose (Promega) in SDS running buffer; store at room temperature. Microwave to melt all the agarose, and allow the solution to cool slightly before use.
14. Front dye solution: 1% (w/v) bromophenol blue (Sigma) in SDS running buffer.

2.4. Deep Purple Staining

1. Fixation solution: 15% (v/v) ethanol; 1% (w/v) citric acid (pH ~2.3) in water.
2. Borate buffer: 100 mM sodium borate in water, pH to 10.5 with NaOH.
3. Staining solution: Right before use, add 1 ml Deep Purple (GE Healthcare) to 500 ml of borate buffer.
4. Washing solution: 15% (v/v) ethanol in water.

3. Methods

3.1. Liquid Culture of Arabidopsis Seedlings

1. Surface sterilize *Arabidopsis* seeds in 40 ml of bleach containing 5.25% sodium hypochlorite for 10 min (see Note 6).
2. Extensively wash the seeds at least five times each with >40 ml sterilized water. Add seeds to flask containing growth medium at a ratio of 50 mg of seeds/250 ml of medium. The growth medium should not exceed 1/4 volume of the flask to ensure enough aeration for the seedlings.
3. Keep imbibed seeds at 4°C for 2 days to synchronize germination, and then move the culture to light chamber to grow under continuous light for 7 days on a shaker at a shaking speed of 90 rpm.
4. Mix all the culture in a large container and split culture mix into two 250-ml beakers for hormone and mock treatment. To minimize sample variation, evenly split seedlings together with old culture medium to avoid sudden osmotic or nutrient shock.

5. For treatment with BL (the most active form of BR), add BL stock solution to 100 nM to one of the beakers to start the treatment, and add the same volume of 85% ethanol to the mock-treated sample. Shake gently for 2 h.
6. Harvest tissue by filtering through a nylon mesh. Use different nylon mesh for BL-treated and mock-treated tissue, or harvest the mock sample first, to avoid cross-BL contamination of the mock sample.

3.2. Purify Plasma Membrane from Liquid-Grown Arabidopsis Seedlings

1. To reduce variations, process BL-treated sample and mock-treated sample in parallel for PM isolation. Reverse the handling order of sample and control in repeat experiments.
2. Homogenize 30–50 g of material (mock-treated sample first) with PowerGen 700D grinder in buffer H (for 1 g tissue, 1–2 ml of grinding buffer should be used) at 10,000 rpm for 5 min. Starting from here, all steps should be performed at 4°C (see Note 7).
3. Filter homogenate through two layers of miracloth, and spin at 10,000 × *g* for 10 min to remove insoluble debris and organelles. Keep the supernatant.
4. Microsomal membranes are pelleted by centrifugation at 60,000 × *g* for 30 min (higher speed or longer spin time makes pellet hard to resuspend).
5. Resuspend pellet in 4–5 ml of buffer R by gently pipetting up and down 150 times. Keep the pipette tip below surface to avoid generating bubbles (see Note 8).
6. On a balance, add 4 g of resuspended microsomal solution (about 20–30 mg microsomal protein) to a 12 g of Dextron T-500/PEG 3350 two-phase mixture (T1) which is prepared the day before PM purification (see Subheading 2). Invert the centrifuge tube 30 times (see Note 9).
7. Spin T1 and 1× buffer R-added T2 at 1,000 × *g*, 4°C, for 5 min to separate the two phases.
8. Remove 8 ml of upper phase from T2 without disturbing the interface to a clean tube for a later balancing purpose. Transfer the same amount (8 ml) of upper phase from T1 to the low phase of T2. Invert the centrifuge tube 30 times to mix solutions.
9. Spin T2 and 1× buffer R-added T3 at 1,000 × *g*, 4°C, for 5 min.
10. Remove 8 ml of upper phase from T3 to a clean tube without disturbing the interface. Transfer the same amount (8 ml) of upper phase from T2 to T3, and invert 30 times.
11. Spin T3 at 1,000 × *g*, 4°C, for 5 min. Transfer 8 ml T3 upper phase from each BR-treated or untreated sample to two to

three clean ultracentrifuge tubes, and dilute with 5–10 volumes of buffer D. Invert ten times to mix.

12. Spin at 100,000 × *g* for 1 h to pellet the PM.
13. Resuspend the PM in 100 μl of buffer D by vortexing (or slowly pipetting up and down without generating bubbles). Alternatively, the PM pellet can be resuspended/solubilized in another buffer suitable for the next analysis step.

3.3. Two-Dimentional Difference Gel Electrophoresis

3.3.1. Label Proteins with CyDye

1. Mix the purified PM in buffer D with five volumes of cold 100% methanol at −20°C overnight to precipitate the protein.
2. Centrifuge at 20,000 × *g* for 15 min to pellet the protein, and discard supernatant.
3. Add 1 ml of −20°C methanol to the pellet, and vortex for 1 min to remove residual salt in the pellet.
4. Centrifuge at 20,000 × *g* for 15 min to pellet the protein, and discard supernatant carefully; the pellet is very loose now.
5. Quick spin and remove residual methanol with a sharp pipette, and immediately add 40 μl 2-DE buffer to the pellet. The pellet should dissolve instantly into the buffer. It is important not to overdry the protein pellet; otherwise, the pellet becomes very difficult to dissolve by the 2-DE buffer in the next step. If the pellet does not dissolve, sonicate at the lowest power for 5–10 s. During sonication, the temperature of 2-DE buffer should be controlled under 37°C. Elevated temperature above 37°C can hydrolyze urea to isocyanate, which modifies proteins by carbamylation.
6. Determine the protein concentration by Bio-Rad protein assay reagent (Bio-Rad) using BSA dissolved in 2-DE buffer as standard.
7. Adjust samples to equal protein concentrations of 3–5 μg/μl with 2-DE buffer.
8. Aliquot 5–10 μl of 2-DE buffer-dissolved PM protein solution, and mix with 0.2 μl of Tris–HCl buffer (1.5 M, pH 8.8) to adjust the pH of the protein solution to around 8.8. Add 50–100 pmol of CyDye (0.5–1 μl of 0.1 mM CyDye working solution) to the protein solution, and incubate the mixture on ice for at least 2 h in the dark for protein labeling (see Note 10).
9. Terminate the labeling reaction by adding 1 μl of 10 mM lysine and incubating on ice for 10 min.

3.3.2. Isoelectric Focusing

1. Combine Cy3- and Cy5-labeled pair of control and treated samples in a 1.5-ml Eppendorf tube.
2. Add 9 μl of 1 M DTT and 4.5 μl of IPG buffer, pH 4–7. Add 100 μg of unlabelled protein each from mock- and BL-treated

sample to make the total protein amount in the mixture around 250–300 μg. Use 2-DE buffer to adjust the final volume of the protein mixture to 450 μl for IEF separation using the 24-cm Immobiline DryStrips (GE Healthcare), pH 4–7 (see Notes 11 and 12).

3. Perform IEF following standard rehydration loading protocol. The typical running conditions are as follows: rehydration for 2 h, active rehydration at 50 V for 10 h, step and hold at 500 V, 1,000 V for 1 h each, and then hold at 8,000 V until reaching a total of 80,000 V-h.

3.3.3. SDS-PAGE

Skip steps 1 and 2 if using precast gels.

1. Treat one of the two glass plates for a gel cassette with Bind Silane. Soak the small low-fluorescence glass plate (the plate without the spacer, GE Healthcare) overnight in laundry bleach, which contains 5.25% sodium hypochlorite, to help scrape off any residual bound gel or CyDye from previous experiments. Wash the plate with Decon solution (GE Healthcare), rinse thoroughly with ddH_2O, and air dry in a dust-free environment. In the chemical hood, lay the dried plate flat on a layer of clean KimWipes. Add 2–3 ml of freshly made Bind-Silane working solution on the plate, and wipe over all areas of the plate with KimWipes or other lint-free cleaning tissue until the solution is completely dry. Cover the plate with KimWipes and leave in the hood for 2 h for excess Bind Silane to evaporate. Wipe clean the Bind Silane-treated surface with 95% ethanol using KimWipes, attach two reference markers on Bind Silane-treated surface, and leave the plate in a dust-free environment for another hour. The reference marker stickers should be placed half way along the edge and 1 cm away from the spacer (see instructions of GE Healthcare for more details).
2. Casting SDS-polyacrylamide gel: Assemble the gel caster according to Ettan DALT electrophoresis unit user's manual (GE Healthcare) using one Bind Silane-treated low-fluorescence glass plate with reference markers and one untreated plate for each gel (if both plates are treated with Bind Silane, you will not be able to open the gel cassette after gel polymerizes). Following standard protocol, prepare 10% SDS-polyacrylamide gel solution using the stock solution listed in sect. 2.3, steps 7–9. Degas the solution with a vacuum pump for 5 min. Add appropriate volume of ammonium persulfate, mix thoroughly by stirring, and pour the acrylamide gel solution into the gel caster until the gel surface is 2 cm below the upper edge of the small glass plate. Add water-saturated isobutanol to cover the acrylamide surface and let the solution polymerize at room temperature for at least 2 h. Best

polymerization can be achieved by letting the gel polymerize overnight at room temperature. Gels can be stored at 4°C for up to 2 weeks; however, best results are achieved using freshly prepared SDS-polyacrylamide gels.

3. At the end of the IEF run, carefully remove the IPG strip from the strip holder and place it in individual equilibration tubes (GE Healthcare) with gel side away from the tube wall (this can be stored in −80°C freezer if the SDS-PAGE gels are not ready). Add 10 ml of SDS equilibration buffer containing 0.5% (w/v) DTT to each tube.
4. Incubate the strips for 15 min with gentle agitation in the dark.
5. Pour off the equilibration buffer in the tube and replace with 10 ml of SDS equilibration buffer containing 4.5% (w/v) iodoacetamide.
6. Incubate the strips for another 15 min with gentle agitation in dark.
7. Briefly rinse the strips by submerging in an equilibration tube containing SDS running buffer.
8. Get the SDS-PAGE gel ready by removing all liquid on top of the gel. Holding one end of the strip with forceps, carefully place the strip in between the two glass plates on top of the SDS-polyacrylamide gel.
9. Add warm agarose overlay solution to cover the strip. Add from one side to avoid trapping of air bubbles. Use a thin flat spatula or spacer to push the strip gently until it reaches the surface of the polyacrylamide gel. Try not to leave any space between the strip and the surface of the polyacrylamide gel, as the space will increase horizontal streaking of gel spots (see Note 13).
10. For each polyacrylamide gel, cut two 5 × 5-mm filter papers. On one filter paper, add 2 μl of PageRuler plus prestained protein ladder (Fermentas). On the other filter paper, add 2 μl of front dye solution. Place the filter papers on top of the SDS-PAGE gel next to each side of the IPG strip (protein ladder on acidic side and the front dye on basic side). Using a spatula or spacer, push the filter papers gently down through the agarose until they touch the surface of the gel.
11. Wait until the agarose solidifies completely, and then start the second dimension SDS gel electrophoresis at 50 V for 2 h, and then at 110 V until the bromphenol blue front reaches the end of the gel (this usually takes 8–12 h).

3.3.4. Gel Scanning, Image Analysis, and Spot Picking

1. At the end of electrophoresis, take out the gel cassettes. Rinse the surface of the glass plates, and scan the gel for Cy3- and Cy5-labeled images using a Typhoon 8600 scanner or a newer model if Cy2 is used (GE Healthcare) with PMT power set at

600. Rescan with different PMT power to obtain images with strong signal without spot signal saturation.

2. Visually examine the image for reddish or greenish spots. Analyze the DIGE image with DeCyder 6.5 software (GE Healthcare) to find the spots that are statistically different between samples and control. Generate a list of spots of interest for gel picking.
3. Open the gel cassette. The SDS-polyacrylamide gel will stick to the Bind Silane-treated glass plate.
4. Fix 2-D DIGE gels in 1,000 ml fixation solution (gel side up) for a minimum of 1.5 h with gentle rocking to avoid scratching the surface of the glass plates. Usually, we do fixation overnight to reduce the background.
5. Move the gels from fixation solution into 500 ml of Deep Purple (GE Healthcare) staining solution (gel side up); try to minimize carryover of the fixation solution. Stain the gels for 1 h in dark with gentle rocking.
6. Wash the gels with 1,000 ml of washing solution for 30 min by gentle rocking.
7. Place the gels in 1,000 ml of fixation solution and rock gently for 30 min. For long-term storage (up to 12 months) of the gels, replace the fixation solution with 1% citric acid containing 1:500 diluted Deep Purple and store the gels at 4°C in the dark.
8. Scan the gel on Typhoon scanner (gel side up) to acquire the Deep purple staining images and Cy5 image. Use Cy5 image as reference gel to mark the spots of interest on Deep Purple image. Use the Decyder software to generate spot picking list.
9. Excise the spots of interest from the gel using an Ettan Spot Picker (GE Healthcare), and process the gel plugs for in-gel digestion and protein identification by mass spectrometry analysis, which can be done by a mass spectrometry facility or service provider.

4. Notes

1. Phosphatase inhibitors (such as cantharidin and sodium orthovanadate) are essential for preventing protein dephosphorylation during sample preparation. Sodium orthovanadate should be activated for maximal inhibition of protein phosphatases. To activate sodium orthovanadate, prepare 200 mM of sodium orthovanadate in ddH_2O and adjust the pH to 10.0 using either 1 N NaOH or 1 N HCl. At pH 10.0, the sodium orthovanadate solution is yellow. Boil the solution until it

turns colorless. Cool to room temperature, and adjust the pH to 10.0 again. Repeat the procedure until the solution remains colorless and the pH stabilizes at pH 10.0. Store the activated sodium orthovanadate as aliquots at −20°C. Add sodium orthovanadate to buffer right before use.

2. In some PM isolation protocols, 0.5% (w/v) Casein and/or 0.5% (w/v) bovine serum albumin (BSA) are included to protect the proteins from degradation after lysis. Since casein and BSA are copurified with the PM vesicles, their relative high abundance creates serious contamination problem later on. In our experience, the casein and/or BSA spots in 2-D DIGE can be used as reference for reproducibility between the samples, but can contaminate other spots in mass spectrometry analysis.
3. Dextran T-500 usually contains 5–10% water, so the concentration of the solution is not 22% but somewhere between 20 and 21%. Exact concentration should be determined with a polarimeter with specific rotation of Dextran at 199° ml/g/dm. If a polarimeter is not available, assuming there is 5% of water in Dextran T-500, large stock solutions (2–3 L) should be prepared so that the same dextran stock is used throughout the whole experiment to ensure reproducible results.
4. For details about the PM purification procedures, readers should read refs. 11, 12. It is advisable to begin with a two-phase mixture with 6.5% polymers and 4–5 mM KCl. Such a two-phase mixture usually yields high-quality PM preparation. Once you are familiar with the methods, you can test different concentrations of polymers and KCl to achieve a good balance between the purity and the yield of the purified PM.
5. The manufacturer recommends using 400 pmol CyDye DIGE Fluor minimal dye (1 μl of 0.4 mM dye stock) to label 50 μg of protein in a small volume (5–10 μl). We found that satisfying 2-D DIGE images can be obtained with as little as 50–100 pmol/30–50 μg of protein in a small volume (5–10 μl), but very clean low-fluorescence glass plates must be used to improve signal-to-background ratio. This practice greatly reduces the cost of 2D-DIGE experiments. To prepare the CyDye working solution, dissolve each 25 nmol CyDye DIGE Fluor minimal dye in 25 μl of freshly opened high-quality anhydrous DMF. Aliquot CyDye solution to a 1.5-ml centrifuge tube (1 μl each) and speed vacuum for 1 min under dim light or until the solution is completely dried. The aliquoted CyDye can now be stored dry at −80°C for at least 6 months without losing observable labeling activity. With prolonged storage time (more than 1 year), we do see decreased labeling efficiency. In that case, double the CyDye labeling concentration can solve the

problem very well. To prepare the 0.1 mM CyDye working solution, add 10 μl DMF to 1 nmol CyDye aliquot, pipette up and down to mix, and use 0.5–1 μl to label 30–50 μg of protein in small volume (5–10 μl). Store the unused CyDye working solution in −80°C. Since the DMF quality is critical for CyDye labeling efficiency, we also aliquot freshly opened high-quality anhydrous DMF in a 1.5-ml centrifuge tube with crew seal cap (1 ml each) and store in −80°C. The aliquoted DMF should be used only once. Once opened again, the remaining DMF should not be reused for future CyDye dissolving, but still can be used as solvent for other less delicate chemicals.

6. Use a 50-ml falcon tube for seeds' sterilization. Sterilizing large amount of seeds in a small-volume centrifuge tube (e.g., 1.5-ml Eppendorf tube) will result in contaminated culture. It is also very important to use the seeds harvested from healthy plants. If the seeds were harvested from fungi-contaminated plants, the surface sterilization methods used in this paper are no longer sufficient.
7. The reason we use a grinder to homogenize the tissue at high speed for a relative long time (5 min) is to get similar yield of microsome for two-phase partition when we start with the equal amount of control and treated tissues.
8. Keep the pipette tip in the solution all the times while gently pipetting up and down. This is a tedious work and requires lots of patience. Once compressed together by centrifugation, it is difficult to separate the membrane vesicle by pipetting. We found that resuspension of microsome directly affects the yield of the PM purification.
9. Vortexing two-phase mixture with microsome greatly increases the PM purity, but significantly decreases the yield.
10. It is found that some proteins are differentially labeled by different CyDyes; therefore, running repeat gels that swap the dyes is crucial to avoid any dye-specific effects. For example, to study BR-regulated PM proteomic changes, we run 2-D DIGE gels for at least five biological repeats (one biological repeat means one experiment start from plant culture). Of these five biological repeats, three of the BR-treated samples were labeled with Cy5, and the other two were labeled with Cy3. Technical repeat of 2-D DIGE of the same protein sample is usually unnecessary unless the first run has obvious technical problems.
11. Usually, we first run at least two biological repeat samples in analytical 2-D DIGE using about 50–100 μg of protein to find the spots of interest, and then run a preparative gel using 500–1,000 μg of protein to pick the interesting spots for in-gel digestion and MS analysis. Low amount of loading (50–100 μg of protein in a 24-cm gel) yields highly resolved

gel images, and higher protein loading decreases the image resolution because increased spot size leads to spot overlap and coverage of low-abundance protein spots by high-abundance spots. However, high loading improves protein identification by mass spectrometry, which is critical for identifying low-abundance proteins. To avoid two separate analytical and preparative gels, a compromise is to load 250–300 μg of protein/gel and pool the corresponding protein spots picked from replicate gels for mass spectrometry analysis. One should keep in mind that the spot signal intensity is largely determined by the amount of CyDye used in 2-D DIGE, and by the amount of protein in post-electrophoresis staining.

12. We recommend using 24-cm IPG strips (pH 4–7) for the initial 2D DIGE analysis. However, narrow pH range IPG strips (e.g., pH 3–5.6, 5.3–6.5, 6–11, and 3.5–4.5) give better separation of spots, and thus more protein spots can be detected and less spot overlap and cross-contamination occur when multiple narrow pH range gels are used rather than a single wide-range gel. For basic protein separation, the 18-cm pH 6–11 IPG strips seem to produce better results than the longer 24-cm pH 7–11 or pH 7–10 IPG strips.
13. Tracking the samples and gel orientation are important through the long procedure. Record in notebook the bar code and number on the IPG strip and the sample loaded on the strip.

Acknowledgment

This work was financially supported by National Science Foundation of China (90917008).

References

1. Tang W, Deng Z, Wang ZY (2010) Proteomics shed light on the brassinosteroid signaling mechanisms. Curr Opin Plant Biol 13:27–33
2. Kosako H, Nagano K (2011) Quantitative phosphoproteomics strategies for understanding protein kinase-mediated signal transduction pathways. Expert Rev Proteomics 8:81–94
3. Choudhary C, Mann M (2010) Decoding signaling networks by mass spectrometry based proteomics. Nat Rev Mol Cell Biol 6:427–439
4. Stasyk T, Huber LA (2004) Zooming in: fractionation strategies in proteomics. Proteomics 4:3704–3716
5. Tang W, Deng Z, Oses-Prieto JA, Suzuki N, Zhu S, Zhang X, Burlingame AL, Wang Z-Y (2008) Proteomics studies of brassinosteroid signal transduction using prefractionation and two-dimensional DIGE. Mol Cell Proteomics 7:728–738
6. O'Farrell PH (1975) High resolution two-dimensional electrophoresis of proteins. J Biol Chem 250:4007–4021
7. Unlu M, Morgan ME, Minden JS (1997) Difference gel electrophoresis: a single gel method for detecting changes in protein extracts. Electrophoresis 18:2071–2077

8. Tonge R, Shaw J, Middleton B, Rowlinson R, Rayner S, Young J, Pognan F, Hawkins E, Currie I, Davison M (2001) Validation and development of fluorescence two-dimensional differential gel electrophoresis proteomics technology. Proteomics 1:377–396
9. Deng Z, Zhang X, Tang W, Oses-Prieto JA, Suzuki N, Gendron JA, Chen H, Guan S, Chalkley RJ, Kaye Peterman T, Burlingame AL, Wang Z-Y (2007) A proteomic study of brassinosteroid response in *Arabidopsis.* Mol Cell Proteomics 6:2058–2071
10. Tang W, Kim T-W, Oses-Prieto JA, Sun Y, Deng Z, Zhu S, Wang R, Burlingame AL, Wang Z-Y (2008) BSKs mediate signal transduction from the receptor kinase BRI1 in *Arabidopsis.* Science 321:557–560
11. Widell S, Larsson C (1987) Plasma membrane purification. In: Senger H (ed) Blue light response: phenomena and occurrence in plants and microorganisms, vol II. CRC Press, Florida
12. Larsson C, Widell S, Kjellbom P (1987) Preparation of high-purity plasma membranes. Methods Enzymol 148:558–568

Chapter 6

Identification and Verification of Redox-Sensitive Proteins in *Arabidopsis thaliana*

Hai Wang, Shengbing Wang, and Yiji Xia

Abstract

Environmental stresses often trigger rapid generation of reactive oxygen species (ROS) in plants. Excessive amount of ROS can cause damage to plant cells and thus need to be counteracted by cellular antioxidant systems. On the other hand, ROS also serve as signaling molecules that modulate various physiological responses and developmental processes. Signaling function of ROS is largely achieved through oxidative modifications of redox-sensitive proteins. Therefore, development of methods for high-throughput identification of redox-sensitive proteins and for verifying and characterizing their in vivo redox states is essential for advancing our understanding of ROS-mediated signaling pathways.

Key words: Proteomics, Redox-sensitive proteins, Oxidative stress, *Arabidopsis thaliana*

1. Introduction

Upon exposure to various environmental stresses, plant cells often rapidly produce high levels of reactive oxygen species (ROS). ROS are not only just toxic by-products of metabolism, but also function as key signaling molecules in physiological responses and developmental processes (1–3). The signaling role of ROS is often carried out through reversible or irreversible oxidation of redox-sensitive proteins, thereby altering their biochemical activities or subcellular localizations. Oxidation–reduction of proteins has increasingly been recognized as an important protein modification mechanism in regulating a wide range of biological processes.

Several proteomics approaches have been developed for detection and identification of oxidant-sensitive thiol-containing proteins (4,5). Figure 1 shows a schematic drawing of a redox proteomics method which is based on differential labeling of oxidized and reduced thiols by a fluorescent tag. Generally, proteins

Zhi-Yong Wang and Zhenbiao Yang (eds.), *Plant Signalling Networks: Methods and Protocols*, Methods in Molecular Biology, vol. 876, DOI 10.1007/978-1-61779-809-2_6,

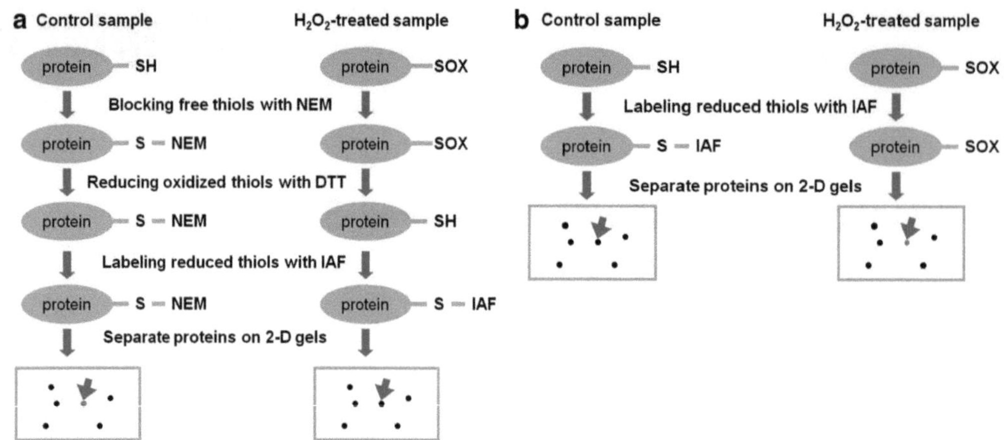

Fig. 1. Schematic drawing of the redox proteomics methods based on 5-iodoacetamidofluorescein (IAF) labeling of protein thiols. (**a**) The blocking method. Proteins were extracted from the control sample and the oxidant-treated sample in the presence of *N*-ethylmaleimide to block free thiols. Oxidized thiols that are present in the samples are reduced and labeled with IAF. After protein separation by two-dimensional electrophoresis (2-DE) and detection through the fluorescence label, differentially labeled protein spots are picked up for mass spectrometry (MS) identification. If the oxidant treatment leads to reversible oxidation of a protein, the protein spot is expected to show an enhanced labeling intensity. (**b**) The direct labeling method. Proteins are extracted from the control sample and the oxidant-treated sample, and free thiols were labeled with IAF. After protein separation by 2-DE and detection, differentially labeled protein spots are picked up for MS analysis. If the oxidant treatment leads to oxidation of a protein, its labeling intensity is expected to be reduced.

are extracted from oxidant-treated and control cells in the presence of a thiol alkylation agent, such as *N*-ethylmaleimide (NEM), to block free thiols (–SH) in samples. Oxidized thiols are reduced with a reducing agent, such as DTT. DTT-reduced thiol groups are then labeled with a thiol-reactive reagent, such as the fluorescent probe 5-iodoacetamidofluorescein (IAF). IAF-labeled fluorescent proteins can then be detected after two-dimensional electrophoresis (2-DE) and protein spots with differential labeling intensity between treated and control samples can be excised and identified by mass spectrometry (MS). The use of the blocking reagent at the initial extraction step serves to prevent redox exchanges of proteins during extraction. This blocking–reducing–labeling method (termed the blocking method here) detects proteins that are reversibly oxidized (reducible by DTT). If certain protein becomes reversibly oxidized by a redox-perturbing treatment, its labeling intensity is expected to increase in the treated sample compared with the control sample. Figure 2 shows two portions of the 2-DE gel images from H_2O_2-treated *Arabidopsis* cells and control cells following protein extraction and IAF labeling using the blocking method.

As an alternative approach, after reducing with DTT, the extracts from control and treated samples can be labeled with spectrally resolvable monofunctional maleimide Cy dyes that react with reduced thiols. For example, Cy5- and Cy3-attached maleimide can be used to label control- and oxidant-treated samples, respectively. After labeling, the protein samples are mixed together and resolved

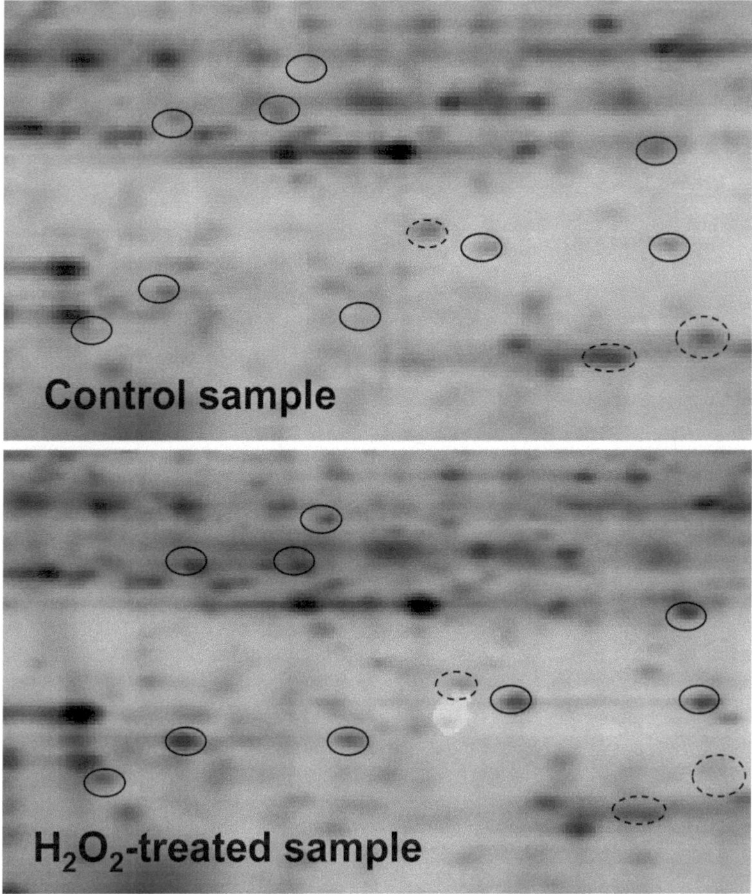

Fig. 2. Detection of differentially oxidized proteins in oxidant-treated *Arabidopsis* cells through the blocking method. Shown are two portions of two-dimensional electrophoresis images of 5-iodoacetamidofluorescein (IAF)-labeled proteins from the oxidant-treated sample (the *lower panel*) and control sample (the *upper panel*). Proteins that were in a more oxidized state (and therefore with a higher label intensity) in the oxidant-treated sample are indicated in the *solid circles*, whereas the spots in *dashed circles* represent proteins in a more reduced state in the oxidant-treated cells. The *Arabidopsis* cells were treated with 5 mM H_2O_2 or H_2O (as a control) for 5 min before being harvested for protein extraction and labeling.

on the same 2-DE gel. This two-dimensional difference gel electrophoresis (2D-DIGE) has been termed redox-DIGE (6).

Protein can also be extracted in the absence of the free thiol-blocking NEM. The protein extract can then be labeled directly with a free thiol-reactive agent, such as IAF. In this method (termed the direct labeling method here) (Fig. 1b), if proteins become oxidized in oxidant-treated samples, they will show a decreased extent of labeling on 2-DE images. Although the direct labeling method is simpler, it detects both reversibly and irreversibly oxidized thiols. Besides, without freezing of redox states by the free thiol-blocking agent, proteins could change their redox states during extraction.

Recently, a modified isotope-coded affinity tag technology (OxICAT) has been developed to identify redox-sensitive proteins in *Escherichia coli* (7). This method uses an isotopically light ICAT

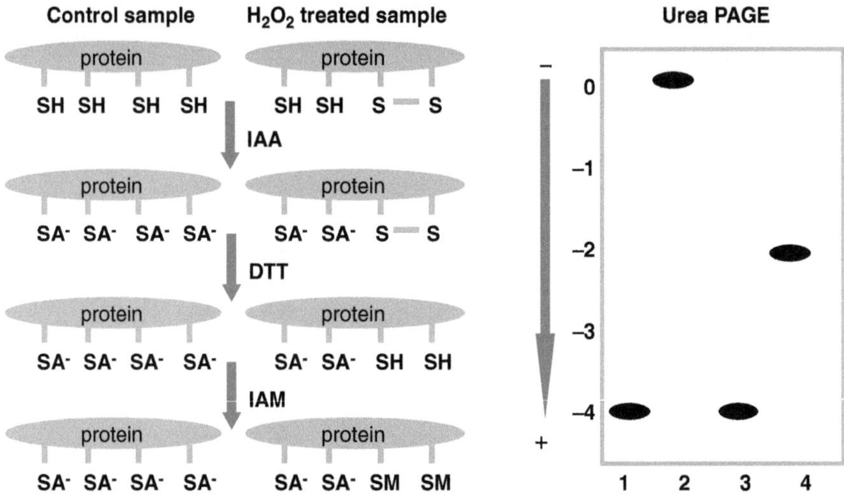

Fig. 3. Schematic drawing of protein electrophoretic mobility shift assay (PEMSA). *Lane 1*: An aliquot of protein crude extract was reduced by DTT and then labeled with iodoacetic acid (IAA). All free thiols and reversibly oxidized thiols are labeled with IAA. *Lane 2*: An aliquot of protein crude extract was reduced by DTT and then labeled with iodoacetamide (IAM). All free thiols and reversibly oxidized thiols are labeled with IAA. The *lanes 2* and *3* represent mobility of the protein when it is IAA and IAM labeled, respectively. *Lanes 3* and *4* represent the protein samples from the control tissue and the H_2O_2-treated tissue following differential labeling of reduced and oxidized thiols with IAA and IAM, respectively.

reagent which consists of the thiol-reactive iodoacetamide (IAM) and a cleavable biotin affinity tag to label reduced thiols and an isotopically heavy ICAT reagent to label reversibly oxidized thiols. In addition to the advantages associated with a liquid-based protein separation and identification, the OxICAT method allows more precise quantification of the extent of oxidative thiol modifications.

These high-throughput proteomics methods aim to identify a large number of redox-sensitive proteins. Methods for verifying individual candidate proteins and detailed characterization of their redox states under various ROS-inducing stress environments are also essential for studying the roles of redox-sensitive proteins in redox signaling. One of such methods is based on protein electrophoretic gel mobility shift assay (PEMSA) (Fig. 3) (8). For this approach, transgenic lines expressing an epitope-tagged candidate protein are generally used for detailed characterization of in vivo redox states of the protein. To determine whether a redox-perturbing condition alters in vivo redox states of the fusion protein in seedlings, transgenic seedlings can be subjected to treatments by oxidants or other redox-perturbing stresses. Proteins are then extracted from the treated and control seedlings. Free thiols are labeled with iodoacetic acid (IAA), and oxidized thiols are then labeled with IAM after being reduced by DTT. IAA carries one negative charge while IAM is neutral. Therefore, the protein that differs in redox states caused by the treatment would have different numbers of negative charges because of differential labeling with

IAA and IAM. This difference could be revealed by separation of the protein by urea-PAGE coupled with Western blot analysis using an antibody against the epitope tag. Proteins with more negative charges migrate relatively faster in a urea gel.

In this chapter, we describe the redox proteomics method based on IAF labeling for identifying redox-sensitive proteins. The basic protocol for protein extraction, blocking free thiols with NEM, and reducing oxidized thiols with DDT can also be applied to other methods, such as redox-DIGE. In addition, we describe the PEMSA method for detailed analysis of in vivo redox states of individual proteins. These methods have been employed in a recent study in the identification and characterization of oxidant-sensitive proteins in Arabidopsis (9).

2. Materials

2.1. A Proteomics Approach to Identify Redox-Sensitive Proteins

2.1.1. Cell Culture and H_2O_2 Treatment

1. *Arabidopsis* suspension culture: T87 suspension culture (RIKEN BioResource Center).
2. Liquid medium for T87 cells: 1/2 MS (Murashige and Skoog) salts, 3% sucrose, 0.29 g/L KH_2PO_4, 0.2 mg/L 2,4-D, and B5 vitamins.
3. 0.5 M H_2O_2 stock solution (prepare just before use).
4. Glass vacuum filter (VWR).

2.1.2. Protein Extraction and Labeling

1. Homogenization buffer: 50 mM Tris–HCl (pH 8.0), 10 mM EDTA, 5 mM EGTA, 50 mM NaCl, 1 mM PMSF, and 1/100 volume of plant protease inhibitor cocktail (Sigma).
2. 10 mM stock of IAF in dimethylformamide (DMF).
3. 1 M stock of NEM in methanol.
4. Phenol (Tris buffered, pH 6.4–6.8).
5. Rehydration buffer: 7 M urea, 2 M thiourea, 4% CHAPS, 20 mM DTT, and 0.5% IPG buffer (Bio-Rad).
6. CB-X Protein Assay kit (G-Biosciences).
7. Reduction buffer: 50 mM Tris–HCl, pH 8.0, 7 M urea, 2 M thiourea, 1% CHAPS, and 1 mM DTT.

2.1.3. Two-Dimensional Gel Electrophoresis

1. 11 cm ReadyStrip IPG strips (pH 4–7) (Bio-Rad).
2. SDS-PAGE running buffer (10×): 250 mM Tris, 1,920 mM glycine, and 1% (w/v) SDS.
3. 8–16% Criterion Precast Gels (Bio-Rad).
4. Sypro Ruby (Bio-Rad).
5. Destaining solution: 10% methanol and 7% acetic acid.

2.1.4. Spot Picking and Mass Spectrometry

1. Multiprobe II Plus (PerkinElmer).
2. Solvent A: 0.1% formic acid in MilliQ water.
3. Solvent B: 0.1% formic acid in ACN.

2.2. Verification of In Vivo Redox Status of Individual Proteins by Protein Electrophoretic Mobility Shift Assay

2.2.1. Plant Growth Medium and Treatment

Liquid 1/2 MS medium: 1/2 MS salts, 5 g/L sucrose, adjusted to pH 5.7 by KOH.

2.2.2. Labeling of YFG-FLAG Fusion Protein with IAA and IAM

1. Urea buffer: 8 M urea, 100 mM Tris–HCl (pH 8.2), 1 mM EDTA.
2. 600 mM IAA stock solution in 1 M Tris–HCl, pH 8.2, prepared just before use.
3. 200 mM IAM stock solution in 1 M Tris–HCl, pH 8.2, prepared just before use.
4. Perfect-FOCUS™ (G-Biosciences).

2.2.3. Urea-PAGE and Western Blotting

1. Running gel: 8 M urea, 9% (w/v) acrylamide, 0.27% (w/v) bisacrylamide, and 0.037 M Tris–HCl (pH 8.8).
2. Stacking gel: 8 M urea, 2.5% (w/v) acrylamide, 0.075% (w/v) bisacrylamide, and 0.12 M Tris–HCl (pH 6.8).
3. 2× Sample buffer: Urea buffer supplemented with 3.5 mM DTT, 16% glycerol, and 0.2% bromphenol blue.
4. Monoclonal ANTI-FLAG® M2-Peroxidase (HRP) antibody (Sigma).

3. Methods

3.1. IAF Labeling-Based Redox Proteomics Approach to Identify Redox-Sensitive Proteins

3.1.1. Cell Culture and H_2O_2 Treatment

1. The T87 cell culture is maintained in 50 ml of liquid medium in 250-ml flasks by gentle agitation (50 rpm) in dark at 22°C. Cells are subcultured weekly by transferring 3 ml of culture to a new flask with 50 ml of the culture medium (see Notes 1 and 2).
2. For the oxidant treatment, a 0.5 M H_2O_2 stock solution is added to T87 cells 3 days after subculturing to a final concentration of 5 mM. For control samples, the same amount of H_2O is added to the culture cells. Ten minutes after the H_2O_2 addition, cells are harvested by filtering through a glass vacuum filter and immediately frozen in liquid nitrogen for further analysis (see Notes 3 and 4).

3.1.2. Protein Extraction and Labeling: For Direct Labeling Method

1. Five hundred microliters of cells are broken in 1.5-ml Eppendorf tubes with plastic micro-pestle on ice in 500 μl homogenization buffer containing 20 μM IAF from a 10 mM stock in DMF and the protein homogenate is incubated on ice for 30 min (see Note 5).
2. The homogenate is then centrifuged for 45 min at 20,000 × *g* under 4°C.
3. The supernatant is mixed with an equal volume of ice-cold phenol and centrifuged at 20,000 × *g* for 15 min at 4°C to separate phenol and aqueous phases.
4. The upper aqueous phase is removed leaving the interface intact, and the phenol phase is extracted twice with 50 mM Tris–HCl, pH 8.0, then mixed with five volumes of cold 0.1 M ammonium acetate in methanol, and left at −20°C overnight to precipitate proteins.
5. After centrifugation at 20,000 × *g* for 15 min, the protein pellet is washed five times with 1 ml of methanol and air dried for 10 min in a fume hood.
6. The pellet is resuspended in the rehydration buffer. After centrifugation at 15,000 × *g* for 3 min, the supernatant is transferred to a new tube, and the protein concentration is determined with the CB-X Protein Assay kit.

3.1.3. Protein Extraction and Labeling: For Blocking Method

1. Five hundred microliters of cells are broken in 1.5-ml Eppendorf tubes with plastic micro-pestle on ice in 500 μl of homogenization buffer containing 10 mM NEM from a 1 M stock in methanol and the protein homogenate is incubated on ice for 30 min (see Note 5).
2. The homogenate is then centrifuged for 45 min at 20,000 × *g* at 4°C.
3. The supernatant is mixed with an equal volume of ice-cold phenol and centrifuged at 20,000 × *g* for 15 min at 4°C to separate phenol and aqueous phases.
4. The upper aqueous phase is removed leaving the interface intact, and the phenol phase is extracted twice with 50 mM Tris–HCl, pH 8.0, then mixed with five volumes of cold 0.1 M ammonium acetate in methanol, and left at −20°C overnight to precipitate proteins.
5. After centrifugation at 20,000 × *g* for 15 min, the protein pellet is washed five times with 1 ml of methanol and air dried for 10 min in a fume hood.
6. The pellet is resuspended in the reduction buffer.
7. After sitting at room temperature for 30 min, 10 mM IAF is added to a final concentration of 40 μM and the mixture is incubated at room temperature for 30 min.

8. The reaction is stopped by adding five volumes of methanol and free IAF is removed by precipitation with centrifugation.
9. The pellet is resuspended in the rehydration buffer (make sure that the concentration of protein exceeds 1.4 μg/μl) and protein concentration is determined with the CB-X Protein Assay kit.

3.1.4. Two-Dimensional Gel Electrophoresis

1. Two hundred and fifty micrograms of protein are mixed with more rehydration buffer to bring the volume to 180 μl.
2. Samples are then applied to 11 cm ReadyStrip IPG strips and the strips are rehydrated overnight at room temperature (see Note 6).
3. IEF is carried out on a Bio-Rad PROTEAN IEF cell at 20°C with maximum 50 μA/strip and the following setting: 250 V for 30 min, 500 V for 1 h, a gradient increase to 8,000 V in 2.5 h, and remaining at 8,000 V until reaching 35,000 V h.
4. After IEF, IPG strips are washed five times by dipping into the SDS-PAGE running buffer.
5. The strips are then transferred to 8–16% Criterion Precast Gels for the second-dimension electrophoresis using Criterion Cell System (Bio-Rad). SDS-PAGE is run at 60 V for 15 min and then at 200 V until the bromphenol blue dye front reaches the gel end (see Note 7).
6. Gels are stained with Sypro Ruby according to the manufacturer's instructions and destained with destaining solution.

3.1.5. Two-Dimensional Gels' Image Scanning and Analysis

1. The gels are scanned using a Typhoon 9410 scanner (GE Healthcare). Signal of IAF-labeled protein is detected using 488-nm laser and a 520-nm band-pass emission filter. Signal of Sypro Ruby fluorescent dyes is detected using 532-nm laser and 610-nm band-pass emission filter. All gels are scanned at 100-μm pixel size (see Note 8).
2. The 2-D gel images are analyzed using Progenesis SameSpots software version 2.0 as described in the user's instruction (Nonlinear Dynamics). Briefly, a sample is chosen as reference image, and alignment vectors are manually added between the reference image and other samples. After alignment, the images of three replicate samples for the same treatment are grouped together and the aligned images are analyzed for spot volume quantification and volume ratio normalization of different samples in the same treatment group. Statistical, quantitative, and qualitative analysis sets are created between the control group and treated group. Protein spots with more than twofold increase or decrease in the normalized spot volume (with p-value <0.05) between the H_2O_2-treated samples and the control samples are picked for identification.

3.1.6. Spot Picking and Mass Spectrometry

1. A spot-picking list generated from Phoretix 2D Evolution gel analysis software (Nonlinear Dynamics Ltd.) is exported to Gelpix (Genetix Inc.). The excised spots are then digested with Multiprobe II Plus.
2. Protein digests are subjected to nano-LC-ESI–MS/MS analysis. Nano-LC is performed with a nanoLC-2D (Eksigent) equipped with a capillary trap LC Packings PepMap (DIONEX) and LC Packings C18 Pep Map 100 (75 μm, 15 cm) connected to the MS. Peptides (5 μl of injections) are desalted for 10 min with a flow rate 5 μl/min of 90.5% solvent A. Peptides are then resolved on a gradient from 9.5 to 35% solvent B for 4 min, from 35 to 45% solvent B for 31 min, and from 45 to 90.5% solvent B over the final 6 min at 200 nl/min flow rate.
3. The MS analysis is performed on an ABI QSTAR XL (Applied Biosystems) hybrid QTOF MS/MS mass spectrometer equipped with a nanoelectrospray source (Protana XYZ manipulator). Positive-mode nanoelectrospray is generated from fused-silica PicoTip emitters with a 10 μm aperture (New Objective) at 2.5 kV. The m/z response of the instrument is calibrated daily with manufacturer standards. TOF mass and product ion spectra are acquired using information-dependent data acquisition (IDA) in Analyst QS v1.1 with the following parameters: mass ranges for TOF MS and MS/MS are m/z 300–2,000 and 70–2,000, respectively. Every second, a TOF MS precursor ion spectrum is accumulated, followed by three product ion spectra, each for 3 s. The switching from TOF MS to MS/MS is triggered by the mass range of peptides (m/z 300–2,000), precursor charge state (2–4), and ion intensity (>50 counts). The DP, DP2, and FP settings are 60, 10, and 230, respectively, and rolling collision energy is used.

3.1.7. Protein Database Search

1. The peptide tandem MS (mass spectra) are processed using Analyst QS software v1.1 (Applied Biosystems) and searched against the NCBI Protein database (5,162,317 sequences, June 2007) using an in-house version of MASCOT v2.20 (Matrix Science Inc).
2. The following parameters are selected: tryptic peptides with ≤1 missed cleavage site; precursor and MS/MS fragment ion mass tolerance of 0.8 and 0.8 Da, respectively; fixed carbamidomethylation of cysteine; and variable oxidation of methionine. Positive identification is determined based on the following criteria: ≥2 peptide sequences; protein sequence coverage; total MASCOT and individual ion scores (http://www.matrixscience.com/help/scoring_help.html); and MS/MS spectral quality judged by a full-length y-ion series of peptides comprising at least six consecutive amino acid sequence tags with no missed cleavages.

3.2. Verification of In Vivo Redox Status of Individual Proteins by Protein Electrophoretic Mobility Shift Assay

3.2.1. Expression of Your Favorite Gene-FLAG Fusion Construct in Arabidopsis

1. Clone Your Favorite Gene (YFG) coding sequence into an appropriate binary vector in which YFG CDS is fused in-frame with a FLAG tag, driven by the 35S promoter (see Notes 9 and 10). The resulting construct is referred to as p35S::YFG-FLAG hereafter.
2. Transform the *Agrobacterium* strain C58C1 with the *35S::YFG-FLAG* construct.
3. Transform wild-type *Arabidopsis* plants by the floral dip method with C58C1 containing *35S::YFG-FLAG* and collect T1 seeds 4 weeks after transformation.
4. Screen T1 plants with appropriate antibiotics according to the selection marker used. Obtain at least ten independent transgenic T1 lines.
5. Find one transgenic T1 plant with detectable expression level of YFP-FLAG fusion protein by Western blot analysis using the monoclonal ANTI-FLAG® M2-Peroxidase (HRP) antibody (Sigma).

3.2.2. Plant Growth and Treatment

1. Put about 20 surface-sterilized homozygous transgenic seeds in 50 ml of liquid 1/2 MS medium in each 200-ml flask. After stratification at 4°C for 3 days, seedlings are grown with gentle (50 rpm) shaking at 22°C under short-day conditions (9/15-h photoperiod with a light intensity of 100 mol/m^2/s).
2. For the H_2O_2 treatment, a 0.5 M H_2O_2 stock solution is added to the flask containing 2-week-old seedlings to a final concentration of 5 mM. For control samples, the same amount of H_2O is added to the flask. Ten minutes after the H_2O_2 addition, plants are harvested and immediately frozen in liquid nitrogen for further analysis (see Note 11).

3.2.3. Labeling of YFG-FLAG Fusion Protein with IAA and IAM

1. 200 μg of plant tissue is homogenized in 1 ml of urea buffer [8 M urea, 100 mM Tris (pH 8.2, 1 mM EDTA) containing 30 mM IAA].
2. Incubate the sample at 37°C for 10 min, and then centrifuge at 15,000 × *g* for 10 min.
3. Purify protein samples using Perfect-FOCUS™ in order to remove residual IAA.
4. Resuspend the final acetone precipitate (you did not mention about acetone precipitation earlier) in 95 μl of urea buffer containing 3.5 mM DTT. And then incubate samples at 37°C for 30 min.
5. Add IAM to a final concentration of 10 mM. Incubate at 37°C for 15 min.
6. For migration markers, 200 μg of seedlings are homogenized in 1 ml of urea buffer containing 3.5 mM DTT. After

incubation at 37°C for 30 min, the samples are divided into two Eppendorf tubes and in each tube, either 30 mM IAA or 10 mM IAM is added. Samples are incubated at 37°C for 15 min.

3.2.4. Urea-PAGE and Western Blotting

1. Cast the running gel and stacking gel as described in Subheading 2.
2. Protein samples were mixed with equal volumes of 2× sample buffer and were loaded into wells.
3. The gel was run at a constant current of 5 mM for an appropriate period of time in the chamber buffer described in Subheading 2.
4. Transfer protein from gel to PVDF or nitrocellulose membrane.
5. Immunoblot detection of YFG-FLAG fusion protein with monoclonal ANTI-FLAG® M2-Peroxidase (HRP) antibody according to the manufacturer's instructions.

4. Notes

1. We used *Arabidopsis* suspension cells rather than whole plant tissues since cells can be exposed to the oxidant more uniformly, which increases the sensitivity for detecting oxidatively modified proteins.
2. Avoid using cell culture older than 3 days for H_2O_2 treatment. Cultures in which cells growing exponentially are better than cultures in the plateau phase which might contain many cells that are undergoing cell death and producing a high level of ROS.
3. We used 0.5–5 mM H_2O_2 to induce oxidative stress of *Arabidopsis* suspension cells and detected similar protein oxidation patterns based on the 2-DE images of IAF-labeled proteins. However, we have not done experiments for direct and extensive comparison on identities of proteins that become oxidatively modified following treatments with different concentrations of the oxidant.
4. H_2O_2 is not very stable and degrades easily. Purchase new H_2O_2 to replace old one every 2 months for reproducible results.
5. All steps of protein extraction, blocking of free thiols, and fluorescence labeling were carried out under reduced light.
6. Choose ready-made strips with appropriate pH ranges according to the isoelectric points of proteins of interest.
7. The size of the gel for the second dimension should be chosen according to the complexity of the protein samples analyzed.

8. When gels were scanned, the photomultiplier tube (PMT) was set to ensure maximum pixel intensity between 40,000 and 80,000 to avoid saturation.
9. Here, we use Flag tag as an example, as the antibody to this epitope tag is commercially available and highly specific. Other epitope tags may be used.
10. Native promoters can also be used if the fusion protein can be readily detected by Western blotting when expressed under the control of the native promoters. Otherwise, a strong promoter, such as the 35S promoter, is preferred.
11. This protocol can be adapted for identification and characterization of oxidatively modified proteins following other types of treatments leading to oxidative modification of proteins, such as salicylic acid, flagellin, plant pathogens, ABA, and abiotic stresses.

References

1. Lamb C, Dixon RA (1997) The oxidative burst in plant disease resistance. Annual Review of Plant Physiology and Plant Molecular Biology 48:251–275
2. Mittler R, Vanderauwera S, Gollery M, Van Breusegem F (2004) Reactive oxygen gene network of plants. Trends Plant Sci 9:490–498
3. Apel K, Hirt H (2004) Reactive oxygen species: metabolism, oxidative stress, and signal transduction. Annu Rev Plant Biol 55:373–399
4. Ghezzi P, Bonetto V (2003) Redox proteomics: identification of oxidatively modified proteins. Proteomics 3:1145–1153
5. Maeda K, Hagglund P, Finnie C, Svensson B (2006) Proteomics of disulphide and cysteine oxidoreduction. In: Finnie C (ed) Plant proteomics. Balckwell Publishing, Oxford, pp 71–97
6. Hurd TR, Prime TA, Harbour ME, Lilley KS, Murphy MP (2007) Detection of reactive oxygen species-sensitive thiol proteins by redox difference gel electrophoresis - Implications for mitochondrial redox signaling. J Biol Chem 282:22040–22051
7. Leichert LI, Gehrke F, Gudiseva HV, Blackwell T, Ilbert M, Walker AK, Strahler JR, Andrews PC, Jakob U (2008) Quantifying changes in the thiol redox proteome upon oxidative stress in vivo. Proc Natl Acad Sci 105:8197–8202
8. Bersani NA, Merwin JR, Lopez NI, Pearson GD, Merrill GF (2002) Protein electrophoretic mobility shift assay to monitor redox state of thioredoxin in cells. Method Enzymol 347:317–326
9. Wang H, Wang S, Lu Y, Alvarez S, Hicks LM, Ge X, Xia Y (2012) Proteomic analysis of early-responsive redox-sensitive Proteins in Arabidopsis. J Proteome Res 11:412–424

Chapter 7

Small-Molecule Dissection of Brassinosteroid Signaling

Mirela-Corina Codreanu, Dominique Audenaert, Long Nguyen, Tom Beeckman, and Eugenia Russinova

Abstract

The growth-promoting hormones, the brassinosteroids (BRs), are perceived at the plant cell surface by receptor kinases that transduce the signal to the nucleus by an intracellular cascade of phosphorylation-mediated protein–protein interactions. BR signaling is also regulated by the plant endocytic machinery because the increased endosomal localization of the BR receptor enhances the BR responses. Chemical genetics is a powerful approach to identify new components in redundant signaling networks and to characterize highly dynamic processes, such as endocytosis. Here, we describe a screen in *Arabidopsis thaliana* seedlings for small molecules that affect hypocotyl elongation under continuous light conditions, indicative for an effect on BR responses. The compounds identified in this screen were used to dissect endomembrane trafficking of the BR receptor, BR INSENSITIVE1, a process that is essential for BR signal transduction.

Key words: *Arabidopsis*, Brassinosteroids, Endocytosis, Chemical genetics, BRI1

1. Introduction

Brassinosteroids (BRs) are growth-promoting hormones with an important role in plant development (1). They act by a direct binding to the plasma membrane (PM)-localized leucine-rich repeat receptor-like kinase BR INSENSITIVE1 (BRI1) and two of its close homologs (2, 3) and subsequently, by modulating the expression of target genes through a signal transduction pathway involving kinases, phosphatases, 14-3-3 proteins, and nuclear transcription factors (4). Growing evidence suggests that the BR signal transduction pathway is regulated through endomembrane trafficking (5–7). BRI1 undergoes ligand-independent endocytosis and recycles back to the PM (5, 6). Increasing the endosomal pool of BRI1 enhances

Zhi-Yong Wang and Zhenbiao Yang (eds.), *Plant Signalling Networks: Methods and Protocols*,
Methods in Molecular Biology, vol. 876, DOI 10.1007/978-1-61779-809-2_7,

BR signaling (6), supporting the hypothesis that endosomal signaling occurs as a continuation of the signaling from the PM.

Genetic studies of BR signaling and its interplay with endocytosis are hampered by gene redundancy, the very dynamic nature of endomembrane trafficking, and the high degree of lethality of the genes encoding endomembrane components (8, 9). Chemical genetics is a powerful approach that overcomes these limitations by the use of small molecules that impair the protein function in a specific, fast, and conditional manner (9). In addition, the function of redundant proteins can be perturbed simultaneously by a general antagonist, thereby revealing novel phenotypes. Additionally, embryonic lethality due to interference with essential signaling components can be avoided by dosage modulation of the chemical compounds or treatment at a later developmental stage (10). Forward chemical genetics screens (from phenotype via chemical to protein) mainly have been applied to study hormonal signaling pathways in plants (11).

Hypocotyl elongation is positively regulated by BRs as illustrated by the short hypocotyl phenotype of the corresponding BR-insensitive or -deficient mutants (12, 13). Chemical screens for hypocotyl growth inhibition in the dark and hypocotyl growth stimulation in the light have identified chemical inhibitors (brassinopride) (14) and activators (bikinin, BIK) of BR responses (15). Whereas the target of brassinopride remains to be identified, BIK inhibits multiple GSK3-like kinases in *Arabidopsis thaliana*, including the major BR-signaling regulator BR INSENSITIVE2 (BIN2) by binding to the ATP-binding pocket and, hence, preventing the phosphorylation of the downstream transcription factor BRI1-EMS-SUPPRESSOR1 (BES1) and triggering BR responses. The usefulness of chemical genetics in analyzing the functional connection between signaling and membrane trafficking can be demonstrated by combining the phenotypic chemical genetic screens for hormonal responses with secondary screens for small chemicals that modulate endomembrane trafficking of signaling components. Here, we present a chemical genetic approach to identify bioactive chemicals that both affect BR responses based on hypocotyl growth (15) and trafficking of BR receptor components based on BRI1 localization in hypocotyls and roots of *Arabidopsis*.

2. Materials

2.1. Equipment

1. Growth chamber.
2. Multiscreen 96-well filter plates (Cat. no. MSBVS1210) (Millipore, Bedford, MA, USA).

3. Square plates (Cat. no. 688102) (Greiner Bio One Labortechnik, Solingen, Germany).
4. Rectangular Nunc dishes with four (Cat. no. 167063) or eight wells (Cat. no. 167064) (Thermo Fisher Scientific, Roskilde, Denmark).
5. Air-permeable tape (Micropore™; 3M, St. Paul, MN, USA).
6. Corning 384-well plates (Cat. no. 3672) (Corning Life Sciences, Lowell, MA, USA).
7. Compact, flat shaker KS 260 basic (IKA Werke GmbH, Staufen, Germany).
8. Vacuum manifold (MSVMHTS00) (Millipore).
9. Vacuum/pressure pump (WP6122050) (Millipore).
10. LightCycler® 480 apparatus (Roche Diagnostics, Brussels, Belgium).
11. Steddy-T Stereo Trinocular Microscope (CETI) (Medline Scientific, Oxford, UK).
12. Confocal microscope Fluoview FV1000 ASW (Olympus, Tokyo, Japan).

2.2. Buffers and Solutions

1. Liquid growth medium: Half-strength Murashige–Skoog (MS) medium supplemented with 1% (w/v) sucrose, 0.01% (w/v) myo-inositol, and 0.05% (w/v) 2-(*N*-morpholino) ethanesulfonic acid (MES) monohydrate. The pH of the medium was adjusted to 5.7 with 1 M KOH. Store MS medium at 4°C.
2. Solid growth medium: MS medium prepared as above and supplemented with 0.8% (w/v) plant tissue culture agar.
3. GUS staining solution: 1 mM 5-bromo-4-chloro-3-indolyl β-D-galactopyranoside sodium salt (X-Glc), 0.5% dimethylformamide (DMF), 0.5% Triton X-100, 1 mM ethylenediaminotetraacetic acid (EDTA) (pH 8), 0.5 mM potassium ferricyanide ($K_3Fe(CN)_6$), and 0.5 mM potassium ferrocyanide ($K_4Fe(CN)_6$). Stock solution of X-Glc, $K_3Fe(CN)_6$ and $K_4Fe(CN)_6$ are kept at −20°C.
4. Phosphate buffer (500 mM): 61.5% of 500 mM K_2HPO_4 and 38.5% of 500 mM KH_2PO_4 (pH 7.0).
5. SYBR Green I Master Kit (Roche Diagnostics).
6. RNeasy Kit (Qiagen, Hilden, Germany).
7. iScript™ cDNA Synthesis Kit (Bio-Rad, Hercules, CA, USA).

2.3. Plant Material

1. For the primary screen, sterilize seeds of *A. thaliana* (L.) Heyhn (ecotype Columbia-0) by treatment with ethanol (70%) for 2 min, followed by NaOCl (15%) for 15 min. Wash five times with sterile water. Stratify seeds in water for 2 days at

4°C. Germinate and grow the sterile and stratified seeds in sterile 96-well filter plates in a growth chamber under continuous light (110 μE/m/s photosynthetically active radiation, supplied by cool-white fluorescent tungsten tubes; Osram, Munich, Germany) at 21°C.

2. For other experiments, germinate the sterile and stratified seeds in square or rectangular plates in a growth chamber under long-day conditions (16-h light/8-h dark) at 21°C.

2.4. Chemicals

1. A 10,000-compound chemical library (DIVERSet™, ChemBridge Corporation, San Diego, CA, USA). All compounds are dissolved in 100% dimethyl sulfoxide (DMSO) at a concentration of 5 mM and stored in 384-well plates at −20°C.
2. DMSO, Brassinolide (BL) (Wako Pure Chemical Industries, Ltd., Tokyo, Japan) and Brassinazole (BRZ) (TCI Europe, Zwijndrecht, Belgium) are used as controls. BL and BRZ were dissolved in 100% DMSO at a concentration of 0.1 and 0.5 mM, respectively, and stored at −20°C.

3. Methods

The methods described below outline the primary phenotypic screen for chemicals affecting hypocotyl growth of *A. thaliana* seedlings in light, the assessment of the effect of the putative hits on BR signaling, the secondary screen for chemicals affecting the localization of BRI1, and the validation strategy for the effect of selected compounds on the BRI1 localization and different endomembrane compartments.

3.1. Chemical Screen for Effectors (Activators or Inhibitors) of Hypocotyl Growth Under Continuous Light Conditions

1. Prepare 96-well filter plates by pipetting 150 μl liquid growth medium under sterile conditions (see Notes 1 and 2).
2. Add manually 7–10 sterilized and stratified seeds of *Arabidopsis* in each well with a pipette.
3. Put plastic lids on the plates and seal with air-permeable tape.
4. Incubate the plates in a growth chamber under continuous shaking at 150 rpm and continuous light to induce germination (see Note 3).
5. After 4 days of seed germination, remove the medium by vacuum filtration with a vacuum manifold and a vacuum/pressure pump and replace with fresh liquid growth medium (see Note 4).
6. Add 1.5 μl of the stock compound to the wells of columns 2–11. Add 1.5 μl of a 100% DMSO solution to the wells of

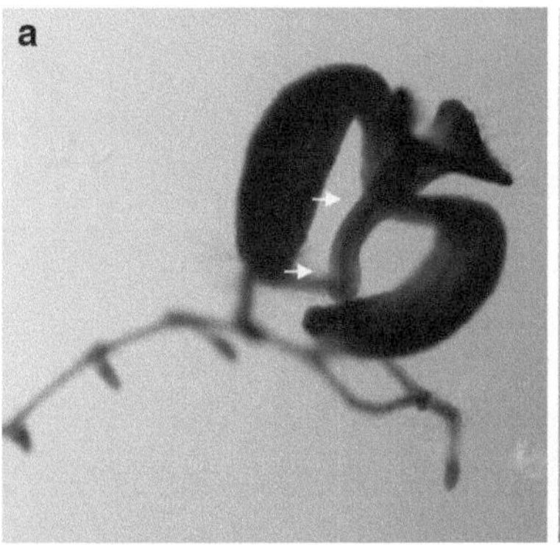

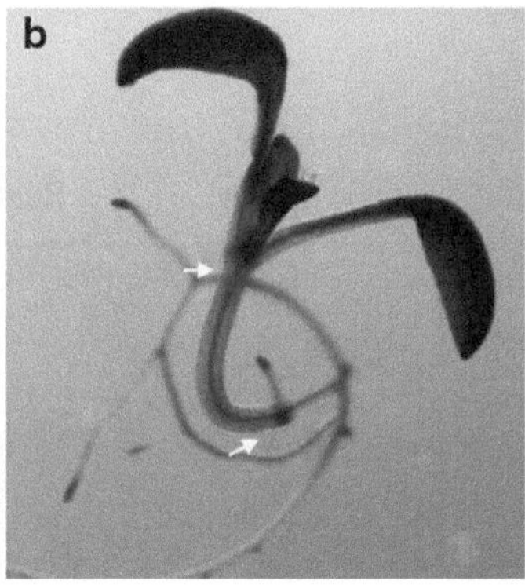

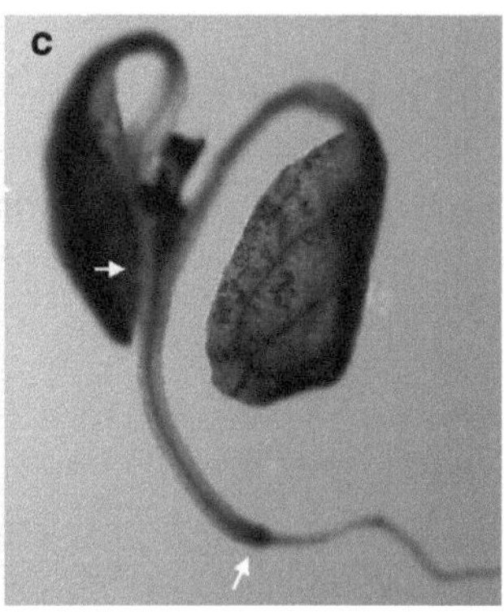

Fig. 1. Phenotypes of 7-day-old seedlings germinated in liquid growth medium supplemented with brassinazole (BRZ) (5 μM) (**a**), dimethyl sulfoxide (DMSO) (**b**), or brassinolide (BL) (1 μM) (**c**). *Arrows* mark both ends of the hypocotyls.

column 12. Add 1.5 μl of the stock BL solution to rows A–D of column 1 and 1.5 μl of the stock BRZ solution to rows E–H of column 1 (see Notes 5 and 6).

7. Incubate the plates in a growth chamber under continuous shaking at 150 rpm and continuous light conditions to allow further growth of the seedlings in the presence of chemicals and control compounds (see Note 3).
8. After 3 days of compound incubation, assess the effect of the chemicals and control compounds on hypocotyl growth under a stereo microscope (see Note 7) (Fig.1).

3.2. Validation of the Primary Screen

1. Reconfirm the effect of every identified hit by growing *Arabidopsis* seeds on solid media supplemented with DMSO and chemicals at a concentration of 1, 3, 5, 10, 25, and 50 μM, respectively (see Note 1).
2. Place the plates vertically in a growth chamber with a photoperiod of 16-h light/8-h darkness.
3. Five days after germination, analyze the length of the primary root and hypocotyl under a stereo microscope (see Note 8).
4. Scan the plates and measure the length of the roots and hypocotyls using ImageJ (available free online; http://www.rsbweb.nih.gov).

3.3. BR Responses

The effect of the hit compounds identified in the primary screen on the BR responses is assessed in different manners.

3.3.1. Quantitative Polymerase Chain Reaction Analysis

By means of a quantitative polymerase chain reaction (qPCR), the expression levels of selected BR biosynthetic signaling and response genes (such as *DWF4*, *CPD*, *ROT3*, *BR6OX1*, *BR6OX2*, *BRI1*, *BIN2*, *BSU1*, *BES1*, *BZR1*, *SAUR-AC1*, *BAS1*, and *NAC*) (15, 16)

Table 1
qPCR primer list

Gene name	Forward primer	Reverse primer
DWF4, At3g50660	GTGATCTCAGCCGTACATTTGGA	CACGTCGAAAAACTACCACTTCCT
CPD, At5g05690	CCCAAACCACTTCAAAGATGCT	GGGCCTGTCGTTACCGAGTT
NAC, At5g46590	GTTTACCTCCAGGGTTCCGGTT	GCACTGAGATGCGACATCTTCG
BAS1, At2g26710	TTGGCTTCATACCGTTTGGC	TTACAGCGAGTGTCAATTTGGC
ROT3, At4g36380	ATTGGCGCGTTCCTCAGAT	CAAGACGCCAAAGTGAGAACAA
BR6OX1, At5g38970	TGGCCAATCTTTGGCGAA	TCCCGTATCGGAGTCTTTGGT
BR6OX2, At3g30180	CAATAGTCTCAATGGACGCAGAGT	AACCGCAGCTATGTTGCATG
BRI1, At4g39400	GGTGAAACAGCACGCAAAACT	CACGCAACCGCAACTTTTAA
BIN2, At4g18710	GTGACTTTGGCAGTGCGAAAC	CAGCATTTTCTCCGGGAAATAATGG
BSU1, At1g03445	GGCGGTTTTCGTCAACAATTCC	CCATCTAAACTGATCTCGGGTAAGG
BES1, At1g19350	CAACCTCGCCTACCTTCAATCTC	TTGGCTGTTCTCAAACTTAAACTCG
BZR1, At1g75080	CCTCTACATTCTTCCCTTTCCTCAG	GCTTAGCGATAGATTCCCAGTTAGG
SAUR-AC1, At4g38850	TTGGGTGCTAAGCAAATTATTCG	TCTCCTACATAGACCGCCATGA
CDKA1;1, At3g48750	ATTGCGTATTGCCACTCTCATAGG	TCCTGACAGGGATACCGAATGC
EEF1α4, At5g60390	CTGGAGGTTTTGAGGCTGGTAT	CCAAGGGTGAAAGCAAG AAGA

are evaluated in the presence of chemicals. The qPCR primers used to analyze the gene expression are given in Table 1.

1. Sow sterilized and stratified *Arabidopsis* seeds on square plates containing solid growth medium (see Note 1).
2. Place the plates vertically in a growth chamber with a photoperiod of 16-h light/8-h darkness.

3. Four days after germination, transfer the seedlings onto 96-well plates containing liquid growth medium and chemicals at a concentration of 1, 3, 5, 10, 25, and 50 μM, respectively. In control treatments, use BL and BRZ at a final concentration of 1 and 3 μM, respectively. The final DMSO concentration in all samples is 1% (see Note 1).
4. Incubate the plates for 2 h in a growth chamber with a photoperiod of 16-h light/8-h darkness.
5. Transfer the seedlings in 2-ml tubes and snap freeze them in liquid nitrogen.
6. Extract RNA with the kit according to the manufacturer's protocol.
7. Synthesize cDNA starting from 1 μg total RNA according to the manufacturer's protocol.
8. Perform the qPCR under the conditions provided by the manufacturer (see Notes 8 and 9).
9. Analyze the results (see Note 10).

3.3.2. Histochemical GUS Analysis

A histochemical GUS analysis is done to allow the visualization of the activation or inhibition of the BR-responsive reporter gene *pCPD:GUS* (17) after compound treatment.

1. Sow sterilized and stratified seeds of the *pCPD:GUS* transgenic line on square plates containing solid growth medium (see Note 1).
2. Place the plates vertically in a growth chamber with a photoperiod of 16-h light/8-h darkness.
3. Four days after germination, transfer ten seedlings to 96-well plates containing liquid growth medium and chemicals at a concentration of 1, 3, 5, 10, 25, and 50 μM, respectively. BL and BRZ are used at final concentrations of 1 and 3 μM, respectively. Final DMSO concentration in all samples is 1% (see Note 1).
4. Incubate the plates for 2 h in a growth chamber with a photoperiod of 16-h light/8-h darkness.
5. Take out the seedlings and incubate them in 90% acetone for 30 min.
6. Wash the seedlings with phosphate buffer.
7. Incubate the seedlings overnight in GUS staining solution at 37°C.
8. Next day, visually identify samples that are darker or lighter blue than the controls (see Note 11).

3.3.3. Phenotypic Analysis of BR Biosynthetic and Signaling Mutants

1. Sow sterilized and stratified seeds of wild-type *Arabidopsis* and mutants (*cpd*, *bri1-9*, *bri1-301*, *bri1-116*, and *bin2-1*) (12, 18, 19) on rectangular plates containing solid growth medium supplemented with chemicals at concentration ranges 1, 3, 5, 10, 25, and 50 μM (see Note 1).
2. Incubate the plates vertically in a growth chamber with a photoperiod of 16-h light/8-h darkness.
3. Four to five days after germination, check visually for phenotypic rescue of the primary root and hypocotyl lengths as in Subheading 3.2 (see Note 8).

3.4. Secondary Screen for Small Chemical Molecules Affecting the Subcellular Localization of BRI1

The effect of the bioactive compounds on the BRI1 localization is monitored by using two transgenic reporter lines stably producing the BRI1 receptor C-terminally fused to the green fluorescent protein (GFP) under the control of the *BRI1* promoter (6, 20).

1. Sow sterilized and stratified seeds on square plates containing solid growth medium (see Note 1).
2. Place the plates vertically in a growth chamber with a photoperiod of 16-h light/8-h darkness.
3. Four days after germination, transfer the seedlings from the square plates into 96-well plates containing liquid growth medium and chemicals at a concentration of 1, 3, 5, 10, 25, and 50 μM, respectively and 1% DMSO as a control.
4. Incubate the plates for 2 h in a growth chamber with a photoperiod of 16-h light/8-h darkness.
5. Take out the seedlings, and mount them on a slide with chemical or DMSO-containing liquid growth medium. Cover the slide with a cover glass.
6. Visualize the meristematic zone of the root tip (Fig. 2) or the hypocotyl epidermis under a confocal microscope with a 63× water-corrected objective (numerical aperture of 1.2 and zoom 4) to scan the cells (see Note 8).
7. Reconfirm the identified hits by imaging BRI1-GFP in roots and hypocotyls treated with chemicals at different concentrations (1, 3, 5, 10, 25, and 50 μM) and DMSO as a control.

3.5. Analysis of the Localization Pattern of Different Endomembrane Markers

To assess in which cellular compartments the chemicals affect the BRI1 trafficking, several fluorescent transgenic lines that mark different endomembrane compartments are used: VHAa1 (21) for Trans-Golgi Network (TGN)/early endosomes, SYP61 (7) for TGN, SYP22 (7) for late endosomes and tonoplast, PIN2 (7) for PM, MAP4 (22) for microtubules, GFP-fABD2 (23) for actin, SYP32-YFP (24) for Golgi apparatus, and NIP1;1-YFP (24) for endoplasmic reticulum and PM. The experiments follow steps described in Subheading 3.4 (see Note 8) (Fig. 2).

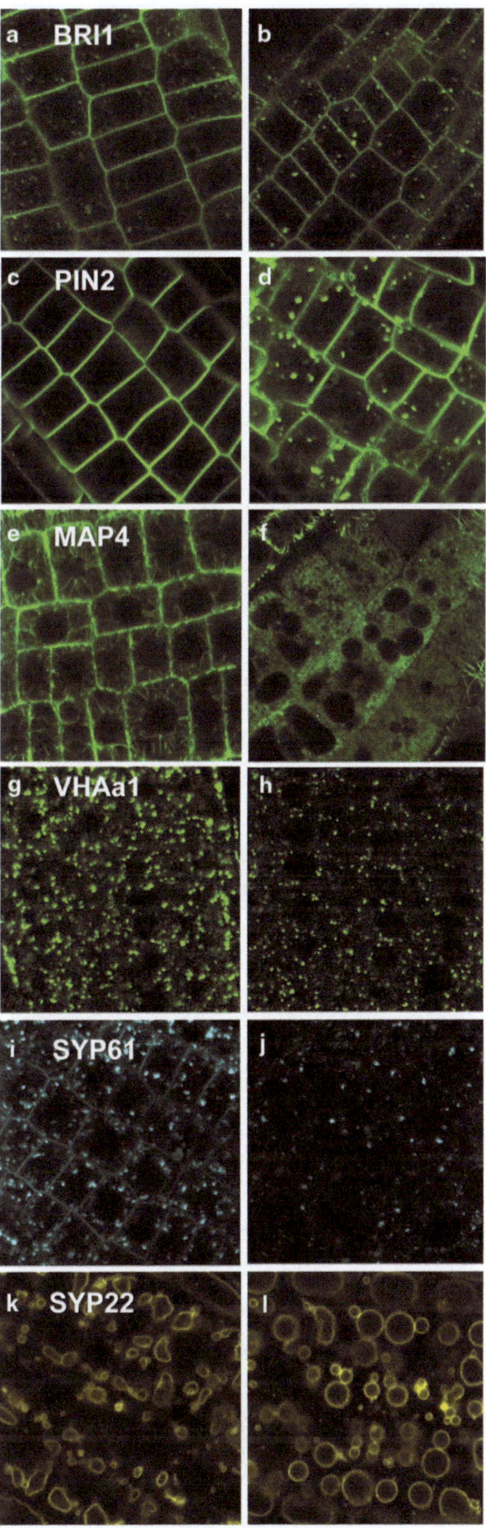

Fig. 2. Observed localization patterns of BRI1-GFP, PIN2-GFP, GFP-MAP4, VHAa1-GFP, SYP61-CFP, and SYP22-YFP in *Arabidopsis* roots after treatment with chemicals (**b**, **d**, **f**, **h**, **j**, and **l**) and dimethyl sulfoxide (DMSO) (**a**, **c**, **e**, **g**, **i**, and **k**).

4. Notes

1. Sterile working conditions should be used to avoid contamination during the germination step.
2. Always use sterile and freshly prepared growth medium.
3. Continuous light and shaking have a positive effect on seed germination and plant growth.
4. During the 4-day germination step and depending on the well position in the microtiter plate, evaporation of the medium will occur. Refreshing the medium assures that the medium levels are the same in all wells and avoids that wells run dry due to further evaporation in the subsequent compound incubation step.
5. This step yields a final DMSO concentration of 1% in all wells, final screening compound concentration of 50 μM, final BL concentration of 1 μM, and final BRZ concentration of 5 μM.
6. To avoid possible inhibition of the germination by some chemicals, add them after the germination step.
7. Before the analysis of the effect of the compounds on the hypocotyl length, one needs to assess whether short hypocotyls (induced by BRZ) can be qualitatively distinguished from long hypocotyls (induced by BL) (Fig. 1).
8. The results must be reconfirmed by repeating the experiment three times. At least ten seedlings should be considered for each compound.
9. Expression levels are normalized to those of *EEF1a4* and *CDKA1;1* (15).
10. Data are analyzed with qBase (25).
11. The histochemical staining pattern and intensity of GUS are examined and chemicals that cause darker staining than the control are marked as BR inhibitors.

Acknowledgments

We thank Karin Schumacher, Jiří Friml, Natasha Raikhel, Niko Geldner, and Daniël Van Damme for providing seeds of marker lines and Martine De Cock for help in preparing the manuscript. This work was supported by a grant from the Research Foundation—Flanders (G.0065.08). DA is a postdoctoral fellow of the Research Foundation—Flanders.

References

1. Clouse SD, Sasse JM (1998) Brassinosteroids: essential regulators of plant growth and development. Annu Rev Plant Physiol Plant Mol Biol 49:427–451
2. Kinoshita T, Caño-Delgado A, Seto H, Hiranuma S, Fujioka S, Yoshida S, Chory J (2005) Binding of brassinosteroids to the extracellular domain of plant receptor kinase BRI1. Nature 433:167–171
3. Caño-Delgado A, Yin Y, Yu C, Vafeados D, Mora-García S, Cheng J-C, Nam KH, Li J, Chory J (2004) BRL1 and BRL3 are novel brassinosteroid receptors that function in vascular differentiation in *Arabidopsis*. Development 131:5341–5351
4. Kim T-W, Wang Z-Y (2010) Brassinosteroid signal transduction from receptor kinases to transcription factors. Annu Rev Plant Biol 61:681–704
5. Russinova E, Borst J-W, Kwaaitaal M, Caño-Delgado A, Yin Y, Chory J, de Vries SC (2004) Heterodimerization and endocytosis of Arabidopsis brassinosteroid receptors BRI1 and AtSERK3 (BAK1). Plant Cell 16:3216–3229
6. Geldner N, Hyman DL, Wang X, Schumacher K, Chory J (2007) Endosomal signaling of plant steroid receptor kinase BRI1. Genes Dev 21:1598–1602
7. Robert S, Chary SN, Drakakaki G, Li S, Yang Z, Raikhel NV, Hicks GR (2008) Endosidin1 defines a compartment involved in endocytosis of the brassinosteroid receptor BRI1 and the auxin transporters PIN2 and AUX1. Proc Natl Acad Sci USA 105:8464–8469
8. Tang W, Deng Z, Wang Z-Y (2010) Proteomics shed light on the brassinosteroid signaling mechanisms. Curr Opin Plant Biol 13:27–33
9. Hicks GR, Raikhel NV (2009) Opportunities and challenges in plant chemical biology. Nat Chem Biol 5:268–272
10. Tóth R, van der Hoorn RAL (2009) Emerging principles in plant chemical genetics. Trends Plant Sci 15:81–88
11. De Rybel B, Audenaert D, Beeckman T, Kepinski S (2009) The past, present and future of chemical biology in auxin research. ACS Chem Biol 4:987–998
12. Szekeres M, Németh K, Koncz-Kálmán Z, Mathur J, Kauschmann A, Altmann T, Rédei GP, Nagy F, Schell J, Koncz C (1996) Brassinosteroids rescue the deficiency of CYP90, a cytochrome P450, controlling cell elongation and de-etiolation in Arabidopsis. Cell 85:171–182
13. Kauschmann A, Jessop A, Koncz C, Szekeres M, Willmitzer L, Altmann T (1996) Genetic evidence for an essential role of brassinosteroids in plant development. Plant J 9:701–713
14. Gendron JM, Haque A, Gendron N, Chang T, Asami T, Wang Z-Y (2008) Chemical genetic dissection of brassinosteroid–ethylene interaction. Mol Plant 1:368–379
15. De Rybel B, Audenaert D, Vert G, Rozhon W, Mayerhofer J, Peelman F, Coutuer S, Denayer T, Jansen L, Nguyen L, Vanhoutte I, Beemster GTS, Vleminckx K, Jonak C, Chory J, Inzé D, Russinova E, Beeckman T (2009) Chemical inhibition of a subset of *Arabidopsis thaliana* GSK3-like kinases activates brassinosteroid signaling. Chem Biol 16:594–604
16. Nemhauser JL, Hong F, Chory J (2006) Different plant hormones regulate similar processes through largely nonoverlapping transcriptional responses. Cell 126:467–475
17. Mathur J, Molnár G, Fujioka S, Takasutso S, Sakurai A, Yokota T, Adam G, Voigt B, Nagy F, Maas C, Schell J, Koncz C, Szekeres M (1998) Transcription of the *Arabidopsis CPD* gene, encoding a steroidogenic cytochrome P450, is negatively controlled by brassinosteroids. Plant J 14:593–602
18. Vert G, Nemhauser JL, Geldner N, Hong F, Chory J (2005) Molecular mechanisms of steroid hormone signaling in plants. Annu Rev Cell Dev Biol 21:177–201
19. Li J, Nam KH (2002) Regulation of brassinosteroid signaling by a GSK3/SHAGGY-like kinase. Science 295:1299–1301
20. Friedrichsen DM, Joazeiro CAP, Li J, Hunter T, Chory J (2000) Brassinosteroid-insensitive-1 is a ubiquitously expressed leucine-rich repeat receptor serine/threonine kinase. Plant Physiol 123:1247–1255
21. Dettmer J, Hong-Hermesdorf A, Stierhof Y-D, Schumacher K (2006) Vacuolar H^+-ATPase activity is required for endocytic and secretory trafficking in *Arabidopsis*. Plant Cell 18:715–730
22. Marc J, Granger CL, Brincat J, Fisher DD, Kao T-h, McCubbin AG, Cyr RJ (1998) A *GFP-MAP4* reporter gene for visualizing cortical microtubule rearrangements in living epidermal cells. Plant Cell 10:1927–1939
23. Sheahan MB, Staiger CJ, Rose RJ, McCurdy DW (2004) A green fluorescent protein fusion to actin-binding domain 2 of Arabidopsis fimbrin highlights new features of a dynamic actin cytoskeleton in live plant cells. Plant Physiol 136:3968–3978

24. Geldner N, Dénervaud-Tendon V, Hyman DL, Mayer U, Stierhof Y-D, Chory J (2009) Rapid, combinatorial analysis of membrane compartments in intact plants with a multicolor marker set. Plant J 59:169–178

25. Hellemans J, Mortier G, De Paepe A, Speleman F, Vandesompele J (2007) qBase relative quantification framework and software for management and automated analysis of real-time quantitative PCR data. Genome Biol 8:R19

Chapter 8

A Chemical Genetics Method to Uncover Small Molecules for Dissecting the Mechanism of ABA Responses in *Arabidopsis* Seed Germination

Yang Zhao

Abstract

Functional redundancy widely exists in genes encoding receptors and signaling components of plant hormones, particularly the stress hormone abscisic acid (ABA). The redundancy hinders the use of conventional genetic approach to dissecting molecular mechanisms for ABA signal perception and transduction. Chemical genetics approach, in which bioactive small molecules are used to perturb the function of gene products encoded by functionally redundant genes, provides an excellent alternative strategy to investigate ABA signaling. This approach led to the discovery of ABA receptor family, PYR/PYL/RCAR (Zhao et al., Nat Chem Biol 3:716–721, 2007; Park et al., Science 324:1068–1071, 2009). A forward chemical genetics screen uncovered an ABA agonist, pyrabactin, which provides a critical small-molecule probe to the identification of the ABA receptor family and the dissection of its downstream signaling.

Key words: Functional redundancy, Abscisic acid, Chemical genetics, Small molecule, Signaling network

1. Introduction

Abscisic acid (ABA) plays multiple important roles in plant growth, development, and stress responses. Under drought stress, ABA crucially regulates stomatal aperture to repress leaf transpiration and prevent plant dehydration (1). Under salinity stress, ABA modulates cell ion homeostasis and activates the accumulation of certain protective molecules to prevent cell damage (2). Recently, ABA was also found to play either positive or negative role in plant disease responses, affecting plants' ability to resist or sensitize certain pathogens (3). ABA also regulates plant development and growth, including embryo and seed development, seedling establishment, and

Zhi-Yong Wang and Zhenbiao Yang (eds.), *Plant Signalling Networks: Methods and Protocols*, Methods in Molecular Biology, vol. 876, DOI 10.1007/978-1-61779-809-2_8,

plant vegetative and reproductive growth. The effect of ABA on seed dormancy and germination has been an easy and important assay for the study of ABA signaling and interaction with other hormones, such as gibberellins (GA), auxin, brassinosteroids, and ethylene. For example, ABA inhibits seed germination, but GA and ethylene counteract the effects of ABA but probably in different phases of seed germination (4). Genetic screens for hypo- or hypersensitivity to ABA inhibition of seed germination have identified a series of genes involved in ABA metabolism and signaling (5). However, no ABA receptors were identified from the screens supposedly due to the functional redundancy of ABA receptor genes and consequently ABA signaling mechanisms remained unclear.

Chemical genetics approaches are complementary to conventional genetics ones. Using functionally active small molecules to conditionally perturb protein function provides chemical genetics approaches some unique advantages in dealing with problems related to genetic redundancy and genetic lethal mutation, which also provides specific perturbations in terms of time and strength. Although applying chemical genetics method to plant science had a shorter history (6) since solid-phase synthetic techniques revolutionized organic molecule synthesis in the second half of the last century and made chemical genetics approach using synthetic molecules to target human diseases feasible (7, 8), it already showed the power in solving problems in plant research of natural genetic variation (9), plant hormone function (10), and the interaction mechanism of plant and its pathogen (11). Currently, the chemical genetics methods have some obvious disadvantages in practice. Firstly, due to the chemical synthesis difficulties, present synthetic molecules provide rather smaller probing chemical spaces than the ones existing in living molecules of organisms. Secondly, in some case, an active small molecule may have affected different proteins' functions of the same organism simultaneously (12).

Perhaps the most successful story of chemical genetics in plant research is a recent forward chemical genetics screen for small molecules mimicking ABA inhibition of *Arabidopsis* seed germination, which led to the discovery of a novel ABA synthetic agonist, pyrabactin, and the follow-up discovery of ABA receptor family, PYR/PYL/RCAR (9, 13). The successful identification of ABA receptor using a chemical genetic-based strategy has proved that chemical genetics provides an extremely valuable alternative approach to investigating functionally redundant genes and complicated signaling networks in plants. This chapter describes a procedure for chemical screen using the identification of pyrabactin as an example.

2. Materials

2.1. Small-Molecule Chemical Library and a Chemical Working Stock Preparation

A novel and structure diverse small-molecule chemical library is purchased from Chembridge Corporation. The library is in a preformatted 96-well plates arranged in the following order: from well *A1* to *H1* and *A12* to *H12*, each contains 100 μl of DMSO (carrier solvent); from well *A2* up to *H11*, each contains 100 μl of 10 mM single chemical stock in DMSO, which is a convenient chemical working stock for using 100 μM as a chemical screen concentration. If a lower chemical screen concentration is required, a corresponding working stock has to be prepared by diluting the original stock using DMSO as solvent. For example, for 25 μM chemical screen concentration, 2.5 mM chemical working stock has to be prepared before the screen. Other similar chemical libraries and various bioactive chemical libraries are available from Chembridge Corporation as well as Life Chemical Corporation.

2.2. Arabidopsis Seeds and Plant Chemical Screen Preparation

1. *Col-0* seeds are purchased from Arabidopsis Biological Resource Center (ABRC, Ohio State University) to restock up to 20 g seeds for a chemical library screen.
2. Seed surface-sterilizing solution: 10% bleach (Clorox), 0.01% Triton X-100 in water, filter sterilized.
3. Seed suspension liquid: 0.1% agar in water, autoclaved.
4. 1× MS solution (Sigma Aldrich).
5. 1% Agar in water, autoclaved.

2.3. High-Throughput Liquid Transaction

Transfer of chemicals from the library and subsequent liquid transfers during plant chemical screen MS plate preparation were carried out using multichannel pipettes (Eppendorf). If possible, an automatic liquid handler or robot is highly recommended for this work.

2.4. Chemical Genetic Phenotype Examination

An optical microscope with an imaging system integrated to a computer-recording workstation (e.g., Leica AF7000, Leica Microsystems) is a least required instrument for examining plant chemical genetic phenotype. To keep a screen in a high-throughput mode, if possible, an automatic high-throughput imaging system for plant phenotype examination is highly recommended, for example Opperetta high throughput imaging system (Perkin Elmer).

3. Method

The chemical screen procedure of this chapter is described as a streamline in Fig. 1.

The screen was carried out in a 96-well format. Each plant screen plate has two rows, including wells from *A1* to *H1* and

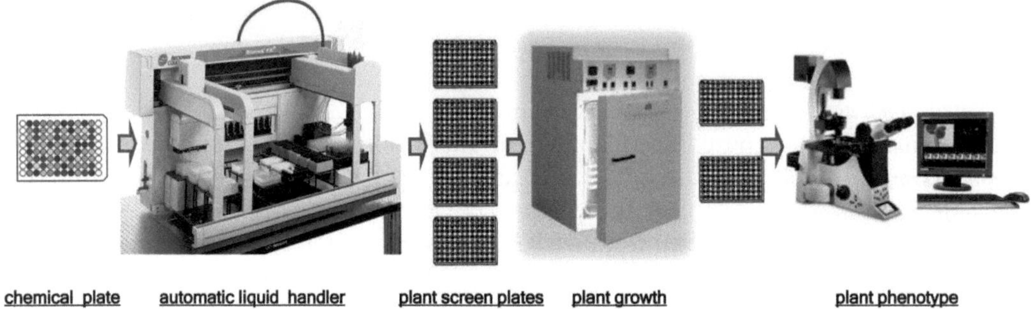

Fig. 1. A *cartoon* description of the steps of a plant chemical genetics screen includes a high-throughput preparation of plant chemical screen plate stage, plant growth stage, and plant chemical genetics phenotype examination stage.

A12 to *H12*, which correspond to wells containing DMSO alone in chemical working stock plates. These wells are for seedlings used as the screen control without chemical treatment.

The screen processes from plant seed preparation, chemical screen MS medium plate preparation to plant phenotype observation take a week. From the second screen week on, for each work day, a screen personnel or team has to finish preparing a certain number of new screen plates and the phenotype examination of a due set of screen plates prepared a week before. Therefore, it should be decided how many new screen plates have to be prepared a day to make sure that the screen personnel or team could complete the plant phenotyping in the same time period.

3.1. Seed Preparation for Chemical Screen

3.1.1. Seed Surface Sterilization

1. For preparing a chemical screen plate, 0.01 g *Arabidopsis Col-0* seeds need to be weighed (see Note 1) into a sterilized Eppendorf or Falcon tube.
2. Add 10× seed volume of seed surface-sterilizing solution into the tube containing seeds, tighten the tube lid, and vortex it thoroughly for 5 s.
3. Keep the tube in room temperature (RT) for 10 min. Spin the tube at 1,000 g in a centrifuge for 1 min at RT and remove the supernatant.
4. Add 20× seed volume of autoclaved distilled water, tight the tube lid, and vortex the seed suspension thoroughly. Spin the tube at 3,000 rpm in a centrifuge for 1 min at RT and pour out the supernatant.
5. Repeat step 4 another 7–10 times until one cannot smell the bleach from the seed pellet.

3.1.2. Prepare Seed Transfer Suspension

1. Add certain volume of seed suspension liquid to the sterilized seed pellet from the previous step (see Note 2): for example, for suspending 0.01 g *Col-0* seeds, use 1.92 ml suspension liquid.

2. Suspend the seed pellet thoroughly through pipetting and vortexing. The final seed suspension should have seed density range from 8 to 12 seeds per 20 μl suspension liquid.
3. The prepared seed suspension can be kept in 4°C refrigerator until use. We have found that this seed suspension could be stored in a 4°C refrigerator for up to 1 week.

3.2. Prepare a Chemical Screen MS Plate

1. For each screen from a chemical library, a duplicate of plant chemical screen MS medium plates needs to be prepared. For this purpose, take two sterilized 96-well plates and label them on one side with the ID number of a chemical working stock plate to be screened.
2. Use a multichannel pipette or automatic liquid handler to transfer 34 μl 1 × MS liquid into each well of aforementioned labeled plates.
3. Use a multichannel pipette or automatic liquid handler to transfer 1 μl chemical stock solution from the chemical working stock plate (see Note 3) into corresponding wells of these labeled plates. For each well from *A1* to *H1* and from *A12* to *H12* of aforementioned labeled plates, instead of 1 μl chemical, 1 μl DMSO from the chemical working stock plate is added.
4. Use a multichannel pipette or automatic liquid handler to transfer 65 μl boiling-hot 1% agar into each well of aforementioned labeled plates (see Note 4). Each plate is kept under RT for solidification for 30 min and then incubated in a 4°C refrigerator for at least 1 h. Now, a pair of plant chemical screen MS medium plates is prepared and ready for use.
5. Use a multichannel pipette or automatic liquid handler to transfer 20 μl seed suspension prepared previously into each well of an MS screen plate (see Note 5).

3.3. Seed Stratification and Germination

1. Chemical screen plates loaded with seed suspension made from the last step are covered with lids and tightly wrapped with cling film. The wrapped plates are placed with lid up in a 4°C refrigerator with a glass door (see Note 6) for 3 days for stratification.
2. When the seed stratification ends, the plates are moved into a dark-grown cabinet (see Note 7) for 4 days to allow seed germination at 22°C.

3.4. Phenotype Examination

1. After the 4-day incubation in the dark, the duplicate screen plates are examined under a microscope for desired phenotypes side by side. A candidate chemical hit is picked up when the same phenotype is observed on both duplicate plates. In the screen for ABA agonists, chemical hits cause *Arabidopsis* seed germination inhibition, as does pyrabactin (Fig. 2).

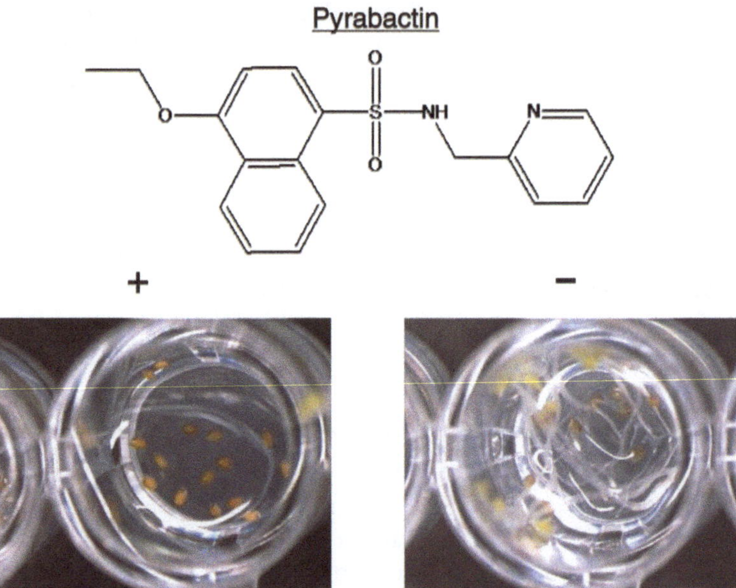

Fig. 2. Pyrabactin inhibits *Arabidopsis Col-0* seeds' germination in a dark-grown condition. The *left photo* shows *Col-0* seeds treated with 25 μM pyrabactin in a dark-grown condition for 4 days; the seeds' germination was inhibited. The *right photo* shows the same batch of *Col-0* seeds grown under the same condition without pyrabactin for 4 days.

2. Seedling phenotypes are imaged and recorded (e.g., Leica AF7000, Leica Microsystems).

3.5. Activity and Structure Characterization of Candidate Chemical Hits

1. For a candidate chemical hit found in a preliminary chemical genetics screen as described previously, a retest needs to be performed to confirm the chemical's biological activity found in the preliminary screen.
2. Upon a candidate chemical hit's biological activity confirmed, a larger amount of the relevant chemical and its various structure analogues need to be ordered.
3. The dose activity measurement and its IC50 have to be determined for the relevant chemical.
4. The dose activity measurement and the IC50 of all structure analogues of the relevant chemical should be determined. This analysis may allow the identification of the chemical's substructure critical for its biological activity and the most active analogue of the relevant chemical.

3.6. Delineate the Mode of Action for a Germination Inhibitory Chemical Comparing with ABA

Now, using the whole genomic microarray experiment to delimit a drug target compass in a model organism has been a routine technique. I will use pyrabactin as an example (13) but not default any practical methodological procedure using a whole genome microarray experiment to determine how similiar a newly uncovered seed germination inhibitory chemical to ABA in term of made of action, since each lab should have its best choice to carry out this kind of experiment.

3.6.1. Determine Dose Effect of ABA and Pyrabactin on Seed Germination Inhibition

Dose activity measurements determine the minimum concentrations of 1 μM ABA and 25 μM pyrabactin for 100% germination inhibition of *Arabidopsis Col-0* seeds grown in dark for 3 days after inhibition (13).

3.6.2. Total RNA Extractions from Chemical-Treated Seeds

For each microarray experiment, RNA extractions from quadruplicates of both ABA and pyrabactin treatments are prepared, meanwhile RNA extractions from triplicate of 1% DMSO treatment are prepared as control. An RNA extraction is prepared from seeds grown on four 150-mm Petri dish plates, each plate grows 2,500 seeds in dark for 24 h on 0.5× MS medium with proper chemical concentration after stratified for 4 days at 4°C (13).

Before RNA extraction, the following buffers and reagents need to be prepared. All chemicals are purchased from Sigma Aldrich.

1. Extraction buffer: 10 mM Tris–HCl (pH 9.5), 10 mM EDTA (pH 8.0), 2% (w/v) lithium dodecyl sulfate, 0.6 M NaCl, 0.4 M trisodium citrate, and 5% (v/v) 2-mercaptoethanol.
2. Phenol mixture: Water-saturated phenol is mixed with 35% (w/v) guanidium thiocyanate and 1/10 volume of 2 M sodium acetate (pH 4.0), stored in dark at 4°C.
3. Chloroform/isoamyl alcohol (v/v: 24:1).
4. Isopropanol.
5. 75% Ethanol (v/v) stored at 4°C.
6. RNase-free water.

RNA Extraction Procedure (Modified from Ref. 14)

1. Collect chemical-treated seeds from growing plates, and grind seeds in powder in liquid nitrogen using a precooled mortar and pestle.
2. Transfer the powder into an ice-bathed 50-ml Falcon tube, add ten volume of extraction of buffer, and mix the mixture by inverting the tightened tube five to seven times. Centrifuge the tube at 14,000 × *g* for 5 min at RT.
3. Transfer every 0.5–0.7 ml supernatant into a set of Eppendorf tubes, add one volume of chloroform/isoamyl alcohol into correspondent tubes, and mix 15 times by inversion. Centrifuge at 14,000 × *g* for 10 min at 4°C. Transfer the upper phase into another set of Eppendorf tubes. From now on, all liquid manipulation is carried out in Eppendorf tube.
4. Add one volume of phenol mixture and mix by inverting tubes ten times. Let tubes to stand for 3 min at RT. Add chloroform/ isoamyl alcohol at 0.5 volume of phenol mixture, shake tubes

vigorously by hand for 20 s, and centrifuge at 14,000 × *g* for 10 min at 4°C.

5. Transfer the upper phase to a set of Eppendorf tubes and add 0.6 volume isopropanol. Mix by inverting tubes 8–10 times and let them stand for 10 min at RT. Centrifuge at 14,000 × *g* for 20 min at 4°C.
6. Wash pellets twice using 75% cold ethanol through centrifugation. Air dry pellets briefly and add 50 μl RNase-free water to dissolve. The RNA solution is stored in −80°C.

Quality and Quantity Examinations of RNA Extractions (13)

For each RNA extraction sample, 1 μl RNA stock solution is diluted into 99 μl 10 mM Tris–HCl (pH 7.4) buffer. The RNA dilution is used to measure RNA concentration and RNA purity through measuring 260 and 280 nm OD ratio which should be between 1.7 and 2.2 by using GeneQuant RNA/DNA Calculator (GE Healthcare Bio-Sciences Corp., NJ, USA). Gel electrophoresis is also employed to examine the states of degradation of the RNA extractions.

3.6.3. Microarray Experiments and Data Analysis (13)

Microarray Experiments (13)

Triplicate 1% DMSO-treated control RNA samples and quadruplicate RNA samples from both ABA and pyrabactin treatments are converted to biotin-labeled cRNA using Enzo kit (Affymetrix, Santa Clara, USA) according to the user instruction of manufacturer and hybridized to 22 k ATH Affymetrix microarray at CAGEF (University of Toronto).

Data Analysis (13)

Gene expression level of microarray is measured by the statistical algorithm for the GCOS/MAS5.0 (Affymetrix). One thousand six hundred and twenty-eight gene probes are identified to response to the treatments of both ABA and pyrabactin comparing with control treatments with fold change and false discovery rate at 5%. Their average transcript levels of both ABA and pyrabactin versus the control ones are transformed in log2 to draw Pearson correlation of mode of action between ABA and pyrabactin treatments on seed germination (Fig. 3).

4. Notes

1. For a certain *Arabidopsis* seed used for a chemical genetics screen, it is a useful practice to find the ratio between the seed weight and its number: for example, for *Col-0* seeds, 1,000 seeds weigh 0.01 g.
2. For a certain *Arabidopsis* seed used for a chemical genetics screen, it is a useful practice to find the ratio between the seed weight and the volume of seed suspension liquid added to

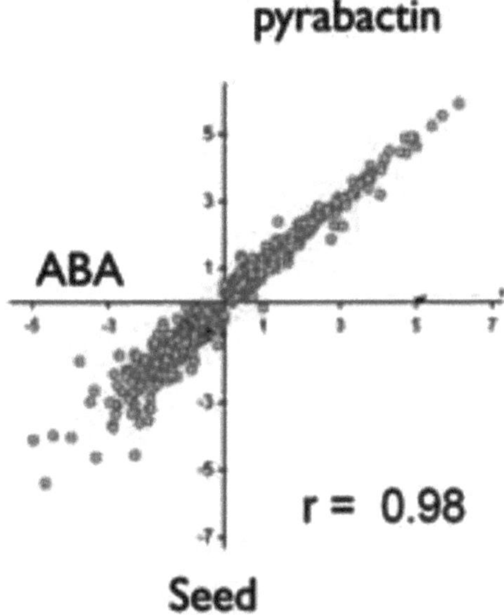

Fig. 3. Pearson correlation of gene expression responses of abscisic acid (ABA) (*x*-axis) and pyrabactin (*y*-axis) treatments versus control ones of *Arabidopsis* seeds' germination indicated an almost the same mode of action of ABA and pyrabactin on seed germination, which showed pyrabactin an ABA functional agonist in seed germination process. Reproduced from ref. 13 with the permission of Science magazine.

acquire a proper seed density in a seed suspension for the seed loading of a chemical screen plate.

3. We have found that less than 2% DMSO in a chemical screen MS medium does not inhibit seed germination and seedling growth in the screen.
4. When adding hot agar liquid into a plate, it is important to push out the hot agar liquid from pipette at a proper speed to prevent any air bubble formed in wells of the plate. In case that some big air bubbles have been formed in some well of the plate, the plate cannot be used for a screen and a new plate needs to be made for the screen.
5. A set of tip-cut pipette tips needs to be prepared for the seed suspension transfer of this step.
6. We have found that when *Arabidopsis* seeds are stratified under a dim light condition at 4°C it helps to break the seed dormancy, for example in a refrigerator with a glass door.
7. We use this dark-grown condition for ABA inhibition of seed germination. Proper growth conditions should be chosen accordingly for specific phenotype assays.

Acknowledgments

The author feels very grateful to have chance to join some very exciting and pioneer works in plant chemical genetics research when studying in Dr. Sean Cutler lab. The author would like to thank Ms. Peiling Zhang from Zhao lab for her kind helping to prepare the manuscript.

References

1. Sirichandra C, Wasilewska A, Vlad F, Valon C, Leung J (2009) The guard cell as a single-cell model towards understanding drought tolerance and abscisic acid action. J Exp Bot 60:1439–1463
2. Verslues PE, Agarwal M, Katiyar-Agarwal S, Zhu J, Zhu J-K (2006) Methods and concepts in quantifying resistance to drought, salt and freezing, abiotic stresses that affect plant water status. Plant J 45:523–539
3. Ton J, Flors V, Mauch-Mani B (2009) The multifaceted role of ABA in disease resistance. Trends Plant Sci 14:310–317
4. Weitbrecht K, Muller K, Leubner-Metzger G (2011) First off the mark: early seed germination. J Exp Bot 62(10):3289–3309
5. Cutler SR, Rodriguez PL, Finkelstein RR, Abrams SR (2010) Abscisic acid: emergence of a core signaling network. Annu Rev Plant Biol 61:651–679
6. Toth R, van der Hoorn RAL (2010) Emerging principles in plant chemical genetics. Trends Plant Sci 15:81–88
7. Sebolt-Leopold JS, Dudley DT, Herrera R, Van Becelaere K, Wiland A, Gowan RC, Tecle H, Barrett SD, Bridges A, Przybranowski S, Leopold WR, Saltiel AR (1999) Blockade of the MAP kinase pathway suppresses growth of colon tumors in vivo. Nat Med 5:810–816
8. Stockwell BR (2000) Chemical genetics: ligand-based discovery of gene function. Nat Rev Genet 1:116–125
9. Zhao Y, Chow TF, Puckrin RS, Alfred SE, Korir AK, Larive CK, Cutler SR (2007) Chemical genetic interrogation of natural variation uncovers a molecule that is glycoactivated. Nat Chem Biol 3:716–721
10. Tsuchiya Y, Vidaurre D, Toh S, Hanada A, Nambara E, Kamiya Y, Yamaguchi S, McCourt P (2010) A small-molecule screen identifies new functions for the plant hormone strigolactone. Nat Chem Biol 6:741–749
11. Schreiber K, Ckurshumova W, Peek J, Desveaux D (2008) A high-throughput chemical screen for resistance to *Pseudomonas syringae* in *Arabidopsis*. Plant J 54:522–531
12. Smukste I, Stockwell BR (2005) Advances in chemical genetics. Annu Rev Genom Hum G 6:261–286
13. Park SY, Fung P, Nishimura N, Jensen DR, Fujii H, Zhao Y, Lumba S, Santiago J, Rodrigues A, Chow TF, Alfred SE, Bonetta D, Finkelstein R, Provart NJ, Desveaux D, Rodriguez PL, McCourt P, Zhu JK, Schroeder JI, Volkman BF, Cutler SR (2009) Abscisic acid inhibits type 2C protein phosphatases via the PYR/PYL family of START proteins. Science 324 (5930):1068–1071
14. Yuji Suzuki TK, Koyama H (2004) RNA isolation from siliques, dry seeds, and other tissues of *Arabidopsis thaliana*. Biotechniques 37:542–544

Chapter 9

Activation Tagging

Xiaoping Gou and Jia Li

Abstract

Insertional mutagenesis is one of the most effective approaches to determine the function of plant genes. However, due to genetic redundancy, loss-of-function mutations often fail to reveal the function of a member of gene families. Activation tagging is a powerful gain-of-function approach to reveal the functions of genes, especially those with high sequence similarity recalcitrant to loss-of-function genetic analyses. Activation tagging randomly inserts a T-DNA fragment containing engineered four copies of enhancer element into a plant genome to activate transcription of flanking genes. We recently generated a new binary vector, *pBASTA-AT2*, which has been efficiently used to discover genes involved in BR biosynthesis, metabolism, and signal transduction. Compared to *pSKI015*, a commonly used activation tagging vector, *pBASTA-AT2*, contains a smaller size of T-DNA and a bigger number of unique restriction sites within the T-DNA region, making cloning of the flanking sequence a lot easier. Our analysis indicated that *pBASTA-AT2* gives dramatically improved transformation efficiency relative to *pSKI015*. In this article, detailed information about this activation tagging vector and the protocol for its application are provided. Three recommended gene cloning approaches based on the use of *pBASTA-AT2*, including inverse PCR, thermal asymmetric interlaced PCR, and adaptor ligation-mediated PCR, are described to identify T-DNA insertion sites after selection of activation-tagged mutant plants.

Key words: Gain-of-function, Activation tagging, pBASTA-AT2, Inverse PCR, Thermal asymmetric interlaced PCR, Adaptor ligation-mediated PCR

1. Introduction

A routinely used approach to dissect a biological process is to generate loss-of-function mutants of genes directly associated with this process. T-DNAs and transposons have been successfully used as tags to generate insertional mutations and identify genes responsible for mutant phenotypes in both model and crop plant species (1–6). Insertional mutagenesis usually produces recessive loss-of-function mutants by introducing T-DNAs or transposons into coding sequences or promoter regions of genes (7). However, insertional

Zhi-Yong Wang and Zhenbiao Yang (eds.), *Plant Signalling Networks: Methods and Protocols*,
Methods in Molecular Biology, vol. 876, DOI 10.1007/978-1-61779-809-2_9,

mutagenesis has its obvious limitations. It is rarely successful to determine roles of genes with functionally redundant homologs. Functional redundancy has been widely identified in almost all sequenced eukaryotic genomes (8). For example, about two-thirds of *Arabidopsis* genes were found to have at least one additional copy in its genome (9). Similarly, many biochemical processes have evolved alternative pathways (10). In addition, some genes are essential to plant growth and development, such as those involved in embryogenesis or gametogenesis. Knocking out these genes could result in lethality. Generating recessive loss-of-function mutants is usually time consuming because the mutant phenotypes can only be examined after homozygous plants are obtained.

Gain-of-function analysis is an alternative and effective approach to unfold function of a gene. This method can overcome the aforementioned limitations of recessive insertional mutagenesis (11). In 1990s, a new concept was developed in the process of screening for genes involved in auxin/cytokinin-independent growth in vitro by Walden's group (12). In this mutagenesis system, a binary vector was constructed by inserting four repetitive enhancer fragments into the T-DNA region. The enhancer sequence was derived from the constitutively active promoter of cauliflower mosaic virus (CaMV) 35S gene. The four copies of enhancer sequence were engineered in the T-DNA near its right border. When the T-DNA is randomly inserted into plant genome by *Agrobacterium*-mediated transformation, it can activate the transcription of gene/genes in the vicinity of the T-DNA insertion site. Because the T-DNA insertion often causes transcriptionally activation of genes, this approach was called activation tagging (13). The original activation tagging vector was successfully used in tissue culture to identify *CKI1*, an important regulatory component of cytokinin signal transduction (14). A large-scale T-DNA-based activation tagging screen in *Arabidopsis* demonstrated that this is an effective approach to generate dominant mutants that are mostly due to overexpression of genes near the T-DNA insertion site (15). This approach has been successfully applied to several species, including Arabidopsis (15–18), rice (19–23), lotus (24), tomato (25), tobacco (26, 27), poplar (28), and petunia (29), to identify genes involved in many aspects of plant growth, such as development (17, 30–36), stress response (37–39), metabolism (26, 27, 40, 41), hormone biosynthesis (29, 42), and catabolism (28, 43). A similar approach using transposable elements has also been used to isolate dominant mutations in Arabidopsis (44–46), rice (47), tobacco (48), barley (49, 50), and lettuce (51). Transposon-based activation tagging is particularly useful in crops that are difficult to transform with T-DNAs (52).

T-DNA-based activation tagging was successfully employed to identify genes involved in brassinosteroid (BR) biosynthesis, homeostasis, and signal transduction (53–59). In this article, a new binary vector for activation tagging is described. A detailed protocol for activation tagging in *Arabidopsis* using this vector is provided.

2. Materials

2.1. Arabidopsis Transformation and Screening

1. *Arabidopsis* plants: Plants of either wild type (Col-0, WS2, etc.) or mutants of your interest (e.g., *bri1-5*, an intermediate mutant allele of *BRI1*, was used to isolate genes involved in BR biosynthesis, metabolism, and signal transduction) can be used for transformation by the activation tagging construct.
2. Activation tagging vector: *pBASTA-AT2* is a binary vector with four tandem repeats of the CaMV 35S enhancer sequence close to its left border (Fig. 1). The mannopine synthase (mas) promoter (Pmas) and the coding region of the glufosinate resistance gene (*BASTA*) were PCR amplified from *pSKI015* (15) and cloned into *pBlueScriptSK*(+) (Stratagene, http://www.genomics.agilent.com). Synonymous mutations were then introduced by site-directed mutagenesis to eliminate all regularly used restriction sites. The cloned sequence was introduced into *pBIB-KAN* (60), and the resulting vector was named *pBASTA*. Four tandem repeats of the CaMV 35S enhancer sequence were subsequently inserted into *pBASTA* to create the *pBASTA-AT2* activation-tagging vector (Fig. 1). The T-DNA region of *pBASTA-AT2* was sequenced, and the resistance of transgenic plants to Basta herbicide was confirmed by spraying with Finale (AgrEvo, Montvale, NJ). The T-DNA region of *pBASTA-AT2* is much smaller than that in *pSKI015* because the *pUC19* sequence in *pSKI015* for plasmid rescue was eliminated in *pBASTA-AT2*. A fragment of about 0.6 kb between the enhancers and the left border facilitates the subsequent cloning.
3. *Agrobacterium* strain: The preferred strain for *Arabidopsis* transformation is GV3101 containing helper plasmid with gentamycin resistance.

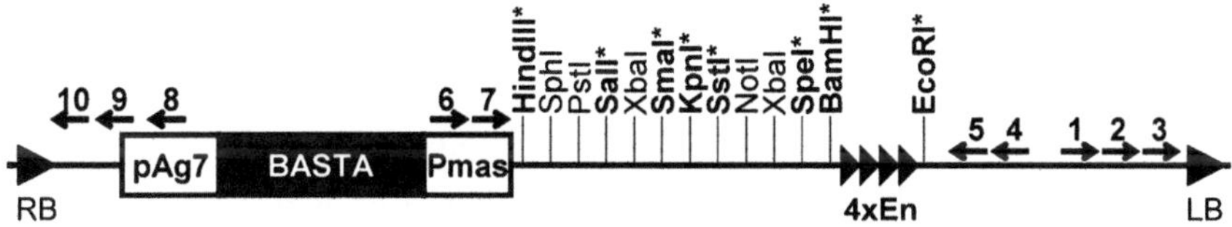

Fig. 1. Schematic representation of the T-DNA of the activation-tagging vector *pBASTA-AT2*. *RB* and *LB* right and left borders of the T-DNA, *Pmas* mannopine synthase (mas) promoter, *BASTA* glufosinate resistance gene, *4 × En* four copies of CaMV 35S enhancers. The T-DNA is in the backbone of pBIB (60). *Asterisks* indicate the unique restriction sites in the vector. *1–10*, primer positions. *1* BIB-TR1, *2* BIB-TR2, *3* BIB-TR3, *4* EX-IPCR1, *5* EX-IPCR2, *6* Pmas5, *7* Pmas6, *8* HB-IPCR3, *9* HB-IPCR4, *10* HB-IPCR5. The primer sequences are listed in Table 1.

4. LB medium: Dissolve 10 g of Bacto Peptone, 5 g of yeast extract, and 5 g of NaCl in 1 L H_2O. Adjust pH to 7.0 with NaOH. Sterilize the medium for 20 min in an autoclave.
5. Antibiotics: Use 25 mg/L gentamycin for the used GV3101 strain and 50 mg/L kanamycin for the binary vector *pBASTA-AT2*.
6. Floral dip medium: 5% sucrose (w/v) and 0.03% (v/v) Silwet L77 (Lehle Seeds, Round Rock, TX, USA).
7. Growth room: Temperature at 20–22°C and 16/8 h of light/dark cycle.
8. Incubator (28°C).

2.2. Determination of Insertion Sites

1. Genomic DNA isolation: Plant DNeasy mini kit (Qiagen, Catalogue No.: 69104).
2. Restriction enzymes: *Bam*H I, *Eco*R I, *Hin*d III, *Kpn* I, *Sac* I, *Sal* I, *Spe* I, and *Xba* I.
3. T4 DNA ligase (1–3 U/μl) with buffer (Promega).
4. Linkers and primers.

   For inverse PCR:

   EX-IPCR1: 5′-GCAACTGTTGGGAAGGGCGATC-3′.

   EX-IPCR2: 5′-AGGGTTTTCCCAGTCACGACGT-3′.

   BIB-TR2: 5′-CAACCCTATCTCGGGCTATTCTT-3′.

   BIB-TR3: 5′-CCGCTTGCTGCAACTCTCTCA-3′.

   HB-IPCR4: 5′-GGAATTGGCTGAGTGGCTCCTT-3′.

   HB-IPCR5: 5′-GTTCCAAACGTAAAACGGCTTG-3′.

   Pmas5: 5′-TTCTCAGACCTACCTCGGCTCT-3′.

   Pmas6: 5′-CGTGAACGGTGAGAAGCTCTG-3′.

   For thermal asymmetric interlaced PCR (TAIL-PCR) (61):

   BIB-TR1: 5′-CCATCGCCCTGATAGACGGTTT-3′.

   BIB-TR2: 5′-CAACCCTATCTCGGGCTATTCTT-3′.

   BIB-TR3: 5′-CCGCTTGCTGCAACTCTCTCA-3′.

   HB-IPCR3: 5′-CATCTACGGCAATGTACCAGCT-3′.

   HB-IPCR4: 5′-GGAATTGGCTGAGTGGCTCCTT-3′.

   HB-IPCR5: 5′-GTTCCAAACGTAAAACGGCTTG-3′.

   AD1: 5′-NTCGASTWTSGWGTT-3′.

   AD2: 5′-NGTCGASWGANAWGAA-3′.

   AD3: 5′-WGTGNAGWANCANAGA-3′.

   For adapter ligation-mediated PCR (62):

   BIB-TR2: 5′-CAACCCTATCTCGGGCTATTCTT-3′.

   BIB-TR3: 5′-CCGCTTGCTGCAACTCTCTCA-3′.

   HB-IPCR4: 5′-GGAATTGGCTGAGTGGCTCCTT-3′.

HB-IPCR5: 5′-GTTCCAAACGTAAAACGGCTTG-3′.

AP1: 5′-GTAATACGACTCACTTAGGGC-3′.

AP2: 5′-TGGTCGACGGCCCGGGCTGC-3′.

LSA: 5′-GTAATACGACTCACTTAGGGCACGCGTGGTC-GACGGCCCGGGCTGC-3′.

SSA *Bam*HI: 5′-phosphate-GATCGCAGCCCG-aminoC7-3′.

SSA *Eco*RI: 5′-phosphate-AATTGCAGCCCG-aminoC7-3′.

SSA *Hind*III: 5′-phosphate-AGCTGCAGCCCG-aminoC7-3′.

5. Incubator (37, 65, and 70°C).
6. Autoclaved tips, Eppendorf tubes, and PCR tubes.
7. Thermocycler.
8. PCR-grade dNTPs (2.5 mM each).
9. ExTaq (5 U/μl) with buffer (TaKaRa), or any other high-efficiency Taq DNA polymerase.

3. Methods

3.1. Plant Growth

Arabidopsis seeds of your choice are treated at 4°C for 3 days in distilled water, then sown in soil, and grown at 21°C under 16/8-h light/dark cycle. Usually, five *Arabidopsis* plants can be grown in a pot of 9 cm side length. Healthy and robust growing plants are the most important factor for efficient transformation of *Arabidopsis*. An appropriate amount of fertilizer should be applied to promote the growth of the plants.

3.2. Agrobacterium-Mediated Arabidopsis Transformation

To screen for desired mutants, large numbers of transgenic plants need to be generated. High transformation efficiency is a key for the success of the approach. Floral dip method (63) is the most suitable transformation approach for *Arabidopsis* activation-tagging screening. The simplified protocol is provided below.

1. Transform *Agrobacterium tumefaciens* strain GV3101 with *pBASTA-AT2* by electroporation or direct transformation into competent cells.
2. When healthy *Arabidopsis* plants begin to produce a mass of flowers, grow a large liquid culture at 28°C in LB medium with 25 mg/L gentamycin and 50 mg/L kanamycin until the culture reaches mid-log or early stationary stage ($OD_{600} = 2.0$). Usually, we inoculate a single colony into 50 ml of LB medium with antibiotics in a 250-ml flask and shake for 30 h.
3. Spin down *Agrobacterium*, and resuspend it in 100 ml of freshly made floral dip medium (see Note 1).
4. Immerse inflorescences of plants into the *Agrobacterium* solution for 10 s with gentle agitation.

5. Cover the dipped plants and keep them from direct illumination of light overnight to maintain high humidity for *Agrobacterium* infection. Plants can be laid on their sides if they grow too tall (see Note 2).

6. Grow and take care of the transformed plants normally and harvest dry seeds after 3–4 weeks.

3.3. Screening for Transgenic Plants

The activation tagging vector *pBASTA-AT2* contains a herbicide selectable marker gene *BASTA* (Fig. 1). The protocol used here usually can produce transgenic plants with efficiency more than 1% (one transgenic plant for every 100 harvested seeds). A trial screening is encouraged to test the true transformation rates. Because mutant phenotypes are screened in the T_1 transgenic plants, it is difficult to screen for mutant phenotypes if too many transgenic plants are grown in a pot. Usually, 20–30 transgenic plants per pot will be the best. Resuspend and wash seeds in distilled water five times and keep at 4°C for 3 days. Sow seeds evenly on top of wet soil using a wide-mouth pipette. After growing for 7 days, spray the seedlings with 0.15% (v/v) commercially available Finale (AgrEvo, Montvale, NJ) containing 11% (w/v) glufosinate in water to screen for transgenic plants with herbicide resistance. Spray again 1 week later to further select the late-germinated seedlings. Screen the population of T_1 transgenic plants by phenotypes of your interest. T_2 seeds from T_1 plants showing no phenotypes can also be collected as pools of 100–500 T_1 plants, and over 400–2,000 T_2 seeds from one pool can be grown and screened for homozygous knockout mutants.

3.4. Genomic DNA Isolation

High quality of genomic DNA is a key to successfully clone the flanking sequence of T-DNA insertion sites. The DNeasy Plant Mini Kit (Qiagen) can be used to purify genomic DNA from activation-tagged *Arabidopsis* mutant plants according to the manufacturer's instruction. An alternative genomic DNA isolation protocol is also provided (see Note 3).

3.5. Determination of Insertion Sites

3.5.1. Inverse PCR

This approach often uses a selected restriction enzyme which cuts the T-DNA only once and cuts the genomic DNA site close to the insertion site. The distance between the genomic cutting site and the T-DNA cutting site differs from one restriction enzyme to another. The larger number of unique sites in the T-DNA region of *pBASTA-AT2* allows selecting for an optimal enzyme for digestion. After intramolecular ligation, the digested DNA fragment can become a circular DNA molecule. The genomic DNA flanking the T-DNA region can, therefore, be amplified by PCR with primers located inversely within the T-DNA region. The T-DNA insertion sites can be determined after DNA sequencing. Detailed protocol to isolate insertion sites of activation tagged lines created by *pBASTA-AT2* is provided.

1. Genomic DNA digestion: In a 100-μl volume, completely digest about 100 ng of purified genomic DNA with one of the following restriction enzymes: *Bam*H I, *Eco*R I, *Hind* III, *Kpn* I, *Sac* I, *Sal* I, *Spe* I, and *Xba* I.
2. Purification of digested genomic DNA: After digestion, stop the reaction by heat inactivation at 70°C for 5 min and purify the reaction with a QIAquick PCR Purification Kit (Qiagen). Elute the purified DNA fragments in 80 μl of sterilized ddH_2O.
3. Self-ligation reaction: Set up the following self-ligation reaction with T4 DNA ligase at 16°C overnight. Less digested DNA used in the reaction allows the intramolecular ligation which is a key for the success of inverse PCR.

Purified digested genomic DNA	80 μl
10× T4 DNA ligase buffer	10 μl
T4 DNA ligase	1 μl
Sterilized ddH_2O	9 μl
Final volume	100 μl

4. Inverse PCR: Purify the ligation reaction with a QIAquick PCR Purification Kit. Elute the purified DNA in 50 μl of sterilized ddH_2O and use 5 μl as template for inverse PCR amplification. The PCR reaction contains the following components:

Purified self-ligated DNA	5 μl
10× ExTaq buffer	2 μl
dNTPs (2.5 mM each)	1.6 μl
Primer 1[a] (10 μM)	0.8 μl
Primer 2[b] (10 μM)	0.8 μl
Sterilized ddH_2O	9.7 μl
ExTaq (5 U/μl)	0.1 μl
Final volume	20 μl

The PCR reaction is performed with 30 cycles (95°C 5 s, 55°C 30 s, 72°C 3 min).

[a]If *Eco*R I or *Xba* I is used to digest genomic DNA, the Primer 1 is EX-IPCR1. If *Bam*H I, *Hind* III, *Kpn* I, *Sac* I, *Sal* I, or *Spe* I is used to digest genomic DNA, the Primer 1 is HB-IPCR3 (Fig. 1, Table 1).

[b]For *Eco*R I- or *Xba* I-digested genomic DNA, the Primer 2 is BIB-TR2. For *Bam*H I, *Hind* III, *Kpn* I, *Sac* I, *Sal* I, or *Spe* I-digested genomic DNA, the Primer 2 is Pmas5 (Fig. 1, Table 1).

Table 1
The primer sequences for the activation-tagging vector *pBASTA-AT2*

Position	Primer name	Primer sequence
1	BIB-TR1	5′-CCATCGCCCTGATAGACGGTTT-3′
2	BIB-TR2	5′-CAACCCTATCTCGGGCTATTCTT-3′
3	BIB-TR3	5′-CCGCTTGCTGCAACTCTCTCA-3′
4	EX-IPCR1	5′-GCAACTGTTGGGAAGGGCGATC-3′
5	EX-IPCR2	5′-AGGGTTTTCCCAGTCACGACGT-3′
6	Pmas5	5′-TTCTCAGACCTACCTCGGCTCT-3′
7	Pmas6	5′-CGTGAACGGTGAGAAGCTCTG-3′
8	HB-IPCR3	5′-CATCTACGGCAATGTACCAGCT-3′
9	HB-IPCR4	5′-GGAATTGGCTGAGTGGCTCCTT-3′
10	HB-IPCR5	5′-GTTCCAAACGTAAAACGGCTTG-3′

Positions of primers in the vector are shown in Fig. 1

5. Gel purification and DNA sequencing: If bright and specific band is obtained, the PCR products can be gel purified and sent for sequencing to determine the insertion site (see Note 4).

3.5.2. Thermal Asymmetric Interlaced PCR

Although inverse PCR often produces specific bands, it requires high-quality genomic DNA, and the whole procedure includes multiple time-consuming enzymatic reactions. An alternative approach is TAIL-PCR. TAIL-PCR was successfully used to recover genomic DNA sequence flanking a T-DNA insertion site (61). Three nested T-DNA-specific primers and one arbitrary degenerate (AD) primer binding to adjacent genomic DNA sequence are utilized to perform three sequential PCR reactions. The T-DNA-specific primers have higher melting temperature (T_m), and the AD primer has lower T_m. Three AD primers (AD1, AD2, and AD3) with 64-, 128-, and 256-fold degeneracy, respectively, can be used (61). Each TAIL-PCR cycle consists of one reduced stringency cycle and two high-stringency cycles, which achieves specific amplification of target sequences. To isolate insertion sites created by *pBASTA-AT2*, two sets of primers are designed for the TAIL-PCR cloning. Primers BIB-TR1, BIB-TR2, and BIB-TR3 locate to the left border of *pBASTA-AT2*. Primers HB-IPCR3, HB-IPCR4, and HB-IPCR5 locate to the right border of the binary vector (Fig. 1).

1. Primary reaction: Set up three reactions (one with each AD primer) as follows:

Component	Amount per well (μl)
Purified genomic DNA (5 ng/μl)	5
10× ExTaq buffer	2
dNTPs (2.5 mM each)	1.6
BIB-TR1 or HB-IPCR3 (10 μM)	0.4
AD1/AD2/AD3 (20/30/40 μM)	2
Sterilized ddH_2O	8.90
ExTaq (5 U/μl)	0.10

Perform the reactions as follows:

Number of cycles	Parameters
1	93°C 1 min, 95°C 1 min
5	94°C 30 s, 62°C 1 min, 72°C 2.5 min
1	94°C 30 s, 25°C 3 min, ramping to 72°C over 2 min, 72°C 2.5 min
20	94°C 10 s, 68°C 1 min, 72°C 2.5 min 94°C 10 s, 68°C 1 min, 72°C 2.5 min 94°C 10 s, 44°C 1 min, 72°C 2.5 min
1	72°C 5 min

2. Secondary reaction: Dilute the three primary reactions 50-fold by transferring 5 μl of each PCR reaction to 245 μl of ddH_2O, respectively. Use 2 μl of each dilution to set up individual secondary reactions as below:

Component	Amount per well (μl)
1:50 Diluted primary PCR reaction	2
10× ExTaq buffer	2
dNTPs (2.5 mM each)	1.6
BIB-TR2 or HB-IPCR4 (10 μM)	0.4
AD1/AD2/AD3 (20/30/40 μM)	2
Sterilized ddH_2O	11.9
ExTaq (5 U/μl)	0.1

Perform the reactions as follows:

Number of cycles	Parameters
18	94°C 10 s, 64°C 1 min, 72°C 2.5 min 94°C 10 s, 64°C 1 min, 72°C 2.5 min 94°C 10 s, 44°C 1 min, 72°C 2.5 min
1	72°C 5 min

3. Tertiary PCR: Dilute the three secondary reactions 50-fold by transferring 5 μl of each PCR reaction to 245 μl of ddH_2O, respectively. Use 2 μl of each dilution to set up individual tertiary reactions as below:

Component	Amount per well (μl)
1:50 Diluted secondary PCR reaction	2
10× ExTaq buffer	2
dNTPs (2.5 mM each)	1.6
BIB-TR3 or HB-IPCR5 (10 μM)	0.4
AD1/AD2/AD3 (20/30/40 μM)	2
Sterilized ddH_2O	11.9
ExTaq (5 U/μl)	0.1

Perform the reactions as follows:

Number of cycles	Parameters
20	94°C 15 s, 44°C 1 min, 72°C 2.5 min
1	72°C 5 min

4. Agarose gel analysis, gel purification, and DNA sequencing: PCR products of all the three reactions for each AD primer are run in adjacent lanes on a 1% agarose gel. The specificity of the reactions can be determined usually by comparing the product sizes between the secondary and tertiary reactions. The primer BIB-TR3 is closer to the LB of *pBASTA-AT2*. The primer HB-IPCR5 is closer to the RB of *pBASTA-AT2* (Fig. 1). The expected tertiary product size will be smaller than that of the secondary reaction. If bright and specific bands are obtained, the PCR products can be gel purified and sent for sequencing using the BIB-TR3 or HB-IPCR5 primers to determine the insertion site.

3.5.3. Adaptor Ligation-Mediated PCR

TAIL-PCR has a tendency to produce false positives because of artifact amplification by PCR with the AD primers. Another cloning strategy was developed to determine insertion sites of T-DNA insertion mutants by using adaptors (62). The adaptors provide specific primers for effective amplification of T-DNA/genomic DNA junction region. Similar to the TAIL-PCR approach, two sets of primers are available for adaptor ligation-mediated PCR amplification of *pBASTA-AT2* insertion sites. Primers BIB-TR2 and BIB-TR3 locate to the left border of *pBASTA-AT2*. Primers HB-IPCR4 and HB-IPCR5 locate to the right border of *pBASTA-AT2* (Fig. 1).

1. Prepare 10× stocks of adaptors. In autoclaved 1.5-ml Eppendorf tubes, mix the following listed oligonucleotides completely by vortexing. Denature the mixtures for 2 min at 96°C by a heat block, and then cool down the heat block to room temperature. Store these 10× stocks at −20°C.

10× BamHI adaptor	10 μl 5 μM LSA 10 μl 5 μM SSA *Bam*HI 605 μl 1 mM Tris, pH 8.3
10× EcoRI adaptor	10 μl 5 μM LSA 10 μl 5 μM SSA *Eco*RI 605 μl 1 mM Tris, pH 8.3
10× HindIII adaptor	10 μl 5 μM LSA 10 μl 5 μM SSA *Hind*III 605 μl 1 mM Tris, pH 8.3

2. Genomic DNA digestion: In a 10-μl volume, completely digest about 100 ng of purified genomic DNA with any of the following restriction enzymes: *Bam*H I, *Eco*R I, and *Hind* III.
3. Purification of digested genomic DNA: After digestion, stop the reaction by heat inactivation at 70°C for 5 min. Add 1 μl of 3 M NaOAc (pH 5.2) and 25 μl of cold 95% (v/v) ethanol into each reaction. Spin down the digested DNA by centrifuging for 10 min at 4°C with full speed (18,000 × *g*). Wash the pellet carefully with 70% ethanol and air dry. Resuspend the pellet in 7.5 μl of autoclaved ddH_2O.
4. Ligase reaction: Add the following components into the tube from the above step to set up the ligase reaction. Keep the tube at 16°C overnight.

Component	Amount per well (μl)
10× T4 DNA ligase buffer	1
10× Stock of adapter	1
T4 DNA ligase (3 U/μl)	0.5

5. Primary adaptor ligation-mediated PCR: Set up the primary PCR reactions by mixing the following components:

Component	Amount per well (μl)
Adaptor-ligated DNA	2
10× ExTaq buffer	2
dNTPs (2.5 mM each)	1.6
BIB-TR2 or HB-IPCR4 (10 μM)	0.5
AP1 (10 μM)	0.5
Sterilized ddH_2O	13.3
ExTaq (5 U/μl)	0.1

Run the following PCR program:

Cycle number	Parameters
1–10	95°C 10 s, 72°C 2.5 min
11–25	95°C 10 s, 62°C 2.5 min

6. Secondary adaptor ligation-mediated PCR: Set up the nested secondary PCR reactions by mixing the following components:

Component	Amount per well (μl)
Primary PCR products	1
10× ExTaq buffer	2
dNTPs (2.5 mM each)	1.6
BIB-TR3 or HB-IPCR5 (10 μM)	0.5
AP2 (10 μM)	0.5
Sterilized ddH_2O	14.3
ExTaq (5 U/μl)	0.1

Run the following PCR program:

Cycle number	Parameters
1–30	95°C 10 s, 62°C 30 s, 72°C 2.5 min

7. Agarose gel analysis, gel purification, and DNA sequencing: PCR products of both the primary and the secondary reactions are run in adjacent lanes on a 1% agarose gel. The specificity of the reactions can be determined usually by comparing the product sizes between the two reactions. The primer BIB-TR3 is closer to the LB of *pBASTA-AT2*. The primer

HB-IPCR5 is closer to the RB of *pBASTA-AT2* (Fig. 1). The expected secondary product size will be smaller than that of the primary reaction. If bright and specific bands are obtained, the PCR products can be gel purified and sent for sequencing using the BIB-TR3 or HB-IPCR5 primers to determine the insertion site.

3.6. RT-PCR Confirmation and Recapitulation

After DNA sequencing, the sequences can be analyzed by BLASTN (64) in TAIR database (http://www.arabidopsis.org) to identify the T-DNA insertion sites. Then, one primer close to the left or right borders in the T-DNA region of *pBASTA-AT2*, such as BIB-TR3 and HB-IPCR5, can be utilized together with another primer from flanking genomic DNA sequence of the possible insertion site to confirm the insertion events by PCR. After the insertion events are confirmed, usually four genes in the vicinity of an insertion site (two upstream and two downstream) are analyzed for their expression level by reverse transcriptase PCR (RT-PCR). If an activated gene or genes can be identified by RT-PCR, recapitulation experiments should be performed by overexpressing the tagged gene or genes under the control of a constitutive promoter, usually the CaMV35S promoter, to test whether the identified gene can reproduce the same phenotypes as in the activation-tagged mutant. In some cases, more than one gene near the T-DNA insertion site can be activated (53). All of the upregulated genes should be examined by recapitulation experiments.

4. Notes

1. Although an *Agrobacterium* culture with $OD_{600} = 0.8$ was used in the original floral dip protocol (63), the authors also pointed out that lower or higher concentration of bacteria culture could give similar transformation rates. To make the procedure even simpler, the bacteria pellet is resuspended in floral dip medium that is two volumes of the original culture for direct use.
2. Healthy plants at optimal growth stage are the key factor for efficient transformation of *Arabidopsis.* Two methods can be used to further improve transformation rate. First, clip primary bolts to produce many secondary bolts and flowers for floral dip. Second, dipping plants two times at 7-day intervals will allow late flowers to be transformed.
3. The DNeasy Plant Mini Kit usually provides *Arabidopsis* genomic DNA with high quality. Therefore, it is suitable for isolating genomic DNA for all the cloning approaches described in this article, especially the restriction enzyme-based cloning approaches, i.e., inverse PCR and adaptor ligation-mediated

PCR. An alternative protocol for genomic DNA isolation is provided for its low cost, which often produces good results. Grind one young leaf (about 0.5–1 cm^2) in liquid nitrogen to fine powder. Add 400 μl extraction buffer [0.1 M Tris–HCl, pH 8.0, 12.5 mM EDTA, 0.5 M NaCl, 1.25% (w/v) SDS, and 10 mM β-mercaptoethanol] into samples and grind until no obvious debris. Keep samples at 65°C for at least 10 min with two to three times of inversion during the process. Add 155 μl of Solution III (3 M potassium acetate and 5 M acetic acid) and mix it completely. Centrifuge for 10 min at 18, 000 × *g* and transfer the supernatant to a new Eppendorf tube containing 250 μl of isopropanol. Mix and store the tubes at −80°C for 30 min. Spin for 15 min at 18, 000 × *g* to pellet isolated genomic DNA. The pellet is then washed once with 70% (v/v) ethanol. The air-dried DNA pellet is then resuspended in 100 μl of autoclaved ddH_2O. The isolated genomic DNA can be stored at −20°C for later use.

4. Usually, a second round of PCR with nested primers provides better PCR results with higher specificity and yield. After the first round of PCR is finished, 1 μl of the 1:50 diluted product of the primary inverse PCR can be used as template for a secondary PCR. The secondary PCR uses the same reaction parameters as the primary PCR with different primer sets. If primers EX-IPCR1 and BIB-TR2 are used for the primary PCR, primers EX-IPCR2 and BIB-TR3 should be used for the secondary PCR. If primers HB-IPCR4 and Pmas5 are used for the primary PCR, primers HB-IPCR5 and Pmas6 should be used for the secondary PCR (Fig. 1, Table 1).

Acknowledgments

We thank Frans Tax for his comments on the manuscript. We are grateful to Yang Ou who designed primers for adaptor ligation-mediated PCR reactions. This study is supported by National Basic Research Program of China Grant 2011CB915401 (to J.L.), and National Natural Science Foundation of China Grants 90917019 (to J.L.) and 31070283 (to X.G.).

References

1. Mathieu M, Winters EK, Kong F et al (2009) Establishment of a soybean (*Glycine max* Merr. L) transposon-based mutagenesis repository. Planta 229:279–289

2. An S, Park S, Jeong DH et al (2003) Generation and analysis of end sequence database for T-DNA tagging lines in rice. Plant Physiol 133:2040–2047

3. Cowperthwaite M, Park W, Xu Z, Yan X, Maurais SC, Dooner HK (2002) Use of the transposon *Ac* as a gene-searching engine in the maize genome. Plant Cell 14:713–726

4. Tadege M, Wen J, He J et al (2008) Large-scale insertional mutagenesis using the *Tnt1* retrotransposon in the model legume *Medicago truncatula*. Plant J 54:335–347
5. Fladung M, Deutsch F, Honicka H, Kumar S (2004) T-DNA and transposon tagging in aspen. Plant Biol (Stuttg) 6:5–11
6. Alonso JM, Stepanova AN, Leisse TJ et al (2003) Genome-wide insertional mutagenesis of *Arabidopsis thaliana*. Science 301:653–657
7. Bouche N, Bouchez D (2001) Arabidopsis gene knockout: phenotypes wanted. Curr Opin Plant Biol 4:111–117
8. Cutler S, McCourt P (2005) Dude, where's my phenotype? Dealing with redundancy in signaling networks. Plant Physiol 138:558–559
9. Arabidopsis Genome Initiative (2000) Analysis of the genome sequence of the flowering plant *Arabidopsis thaliana*. Nature 408:796–815
10. Gu Z, Steinmetz LM, Gu X, Scharfe C, Davis RW, Li WH (2003) Role of duplicate genes in genetic robustness against null mutations. Nature 421:63–66
11. Kondou Y, Higuchi M, Matsui M (2010) High-throughput characterization of plant gene functions by using gain-of-function technology. Annu Rev Plant Biol 61:373–393
12. Hayashi H, Czaja I, Lubenow H, Schell J, Walden R (1992) Activation of a plant gene by T-DNA tagging: auxin-independent growth *in vitro*. Science 258:1350–1353
13. Walden R, Fritze K, Hayashi H, Miklashevichs E, Harling H, Schell J (1994) Activation tagging: a means of isolating genes implicated as playing a role in plant growth and development. Plant Mol Biol 26:1521–1528
14. Kakimoto T (1996) CKI1, a histidine kinase homolog implicated in cytokinin signal transduction. Science 274:982–985
15. Weigel D, Ahn JH, Blazquez MA et al (2000) Activation tagging in Arabidopsis. Plant Physiol 122:1003–1013
16. Robinson SJ, Tang LH, Mooney BA et al (2009) An archived activation tagged population of *Arabidopsis thaliana* to facilitate forward genetics approaches. BMC Plant Biol 9:101
17. Nakazawa M, Ichikawa T, Ishikawa A, Kobayashi H, Tsuhara Y, Kawashima M, Suzuki K, Muto S, Matsui M (2003) Activation tagging, a novel tool to dissect the functions of a gene family. Plant J 34:741–750
18. Ichikawa T, Nakazawa M, Kawashima M et al (2003) Sequence database of 1172 T-DNA insertion sites in Arabidopsis activation-tagging lines that showed phenotypes in T1 generation. Plant J 36:421–429
19. Hsing YI, Chern CG, Fan MJ et al (2007) A rice gene activation/knockout mutant resource for high throughput functional genomics. Plant Mol Biol 63:351–364
20. Wan S, Wu J, Zhang Z et al (2009) Activation tagging, an efficient tool for functional analysis of the rice genome. Plant Mol Biol 69:69–80
21. Jeong DH, An S, Park S et al (2006) Generation of a flanking sequence-tag database for activation-tagging lines in japonica rice. Plant J 45:123–132
22. Lee YS, Jeong DH, Lee DY et al (2010) OsCOL4 is a constitutive flowering repressor upstream of Ehd1 and downstream of OsphyB. Plant J 63:18–30
23. Jeong DH, An S, Kang HG, Moon S, Han JJ, Park S, Lee HS, An K, An G (2002) T-DNA insertional mutagenesis for activation tagging in rice. Plant Physiol 130:1636–1644
24. Imaizumi R, Sato S, Kameya N, Nakamura I, Nakamura Y, Tabata S, Ayabe S, Aoki T (2005) Activation tagging approach in a model legume, *Lotus japonicus*. J Plant Res 118:391–399
25. Mathews H, Clendennen SK, Caldwell CG et al (2003) Activation tagging in tomato identifies a transcriptional regulator of anthocyanin biosynthesis, modification, and transport. Plant Cell 15:1689–1703
26. Ahad A, Nick P (2007) Actin is bundled in activation-tagged tobacco mutants that tolerate aluminum. Planta 225:451–468
27. Ahad A, Wolf J, Nick P (2003) Activation-tagged tobacco mutants that are tolerant to antimicrotubular herbicides are cross-resistant to chilling stress. Transgenic Res 12:615–629
28. Busov VB, Meilan R, Pearce DW, Ma C, Rood SB, Strauss SH (2003) Activation tagging of a dominant gibberellin catabolism gene (GA 2-oxidase) from poplar that regulates tree stature. Plant Physiol 132:1283–1291
29. Zubko E, Adams CJ, Machaekova I, Malbeck J, Scollan C, Meyer P (2002) Activation tagging identifies a gene from *Petunia hybrida* responsible for the production of active cytokinins in plants. Plant J 29:797–808
30. Kardailsky I, Shukla VK, Ahn JH, Dagenais N, Christensen SK, Nguyen JT, Chory J, Harrison MJ, Weigel D (1999) Activation tagging of the floral inducer FT. Science 286:1962–1965
31. van der Graaff E, Dulk-Ras AD, Hooykaas PJ, Keller B (2000) Activation tagging of the *LEAFY PETIOLE* gene affects leaf petiole development in *Arabidopsis thaliana*. Development 127:4971–4980
32. Huang S, Cerny RE, Bhat DS, Brown SM (2001) Cloning of an Arabidopsis patatin-like

gene, *STURDY*, by activation T-DNA tagging. Plant Physiol 125:573–584

33. van der Graaff E, Hooykaas PJ, Keller B (2002) Activation tagging of the two closely linked genes *LEP* and *VAS* independently affects vascular cell number. Plant J 32:819–830
34. Zuo J, Niu QW, Frugis G, Chua NH (2002) The *WUSCHEL* gene promotes vegetative-to-embryonic transition in Arabidopsis. Plant J 30:349–359
35. Woodward C, Bemis SM, Hill EJ, Sawa S, Koshiba T, Torii KU (2005) Interaction of auxin and ERECTA in elaborating Arabidopsis inflorescence architecture revealed by the activation tagging of a new member of the YUCCA family putative flavin monooxygenases. Plant Physiol 139:192–203
36. Xu YY, Wang XM, Li J, Li JH, Wu JS, Walker JC, Xu ZH, Chong K (2005) Activation of the *WUS* gene induces ectopic initiation of floral meristems on mature stem surface in *Arabidopsis thaliana*. Plant Mol Biol 57:773–784
37. Yu H, Chen X, Hong YY et al (2008) Activated expression of an Arabidopsis HD-START protein confers drought tolerance with improved root system and reduced stomatal density. Plant Cell 20:1134–1151
38. Aboul-Soud MA, Chen X, Kang JG, Yun BW, Raja MU, Malik SI, Loake GJ (2009) Activation tagging of *ADR2* conveys a spreading lesion phenotype and resistance to biotrophic pathogens. New Phytol 183:1163–1175
39. Koiwa H, Bressan RA, Hasegawa PM (2006) Identification of plant stress-responsive determinants in Arabidopsis by large-scale forward genetic screens. J Exp Bot 57:1119–1128
40. Borevitz JO, Xia Y, Blount J, Dixon RA, Lamb C (2000) Activation tagging identifies a conserved MYB regulator of phenylpropanoid biosynthesis. Plant Cell 12:2383–2394
41. van der Fits L, Hilliou F, Memelink J (2001) T-DNA activation tagging as a tool to isolate regulators of a metabolic pathway from a genetically non-tractable plant species. Transgenic Res 10:513–521
42. Sun J, Niu QW, Tarkowski P, Zheng B, Tarkowska D, Sandberg G, Chua NH, Zuo J (2003) The Arabidopsis *AtIPT8/PGA22* gene encodes an isopentenyl transferase that is involved in de novo cytokinin biosynthesis. Plant Physiol 131:167–176
43. Schomburg FM, Bizzell CM, Lee DJ, Zeevaart JA, Amasino RM (2003) Overexpression of a novel class of gibberellin 2-oxidases decreases gibberellin levels and creates dwarf plants. Plant Cell 15:151–163
44. Kirch T, Simon R, Grunewald M, Werr W (2003) The *DORNROSCHEN/ENHANCER OF SHOOT REGENERATION1* gene of Arabidopsis acts in the control of meristem ccll fate and lateral organ development. Plant Cell 15:694–705
45. Schneider A, Kirch T, Gigolashvili T, Mock HP, Sonnewald U, Simon R, Flugge UI, Werr W (2005) A transposon-based activation-tagging population in *Arabidopsis thaliana* (TAMARA) and its application in the identification of dominant developmental and metabolic mutations. FEBS Lett 579:4622–4628
46. Marsch-Martinez N, Greco R, Van Arkel G, Herrera-Estrella L, Pereira A (2002) Activation tagging using the *En-I* maize transposon system in Arabidopsis. Plant Physiol 129:1544–1556
47. An G, Jeong DH, Jung KH, Lee S (2005) Reverse genetic approaches for functional genomics of rice. Plant Mol Biol 59:111–123
48. Suzuki Y, Uemura S, Saito Y, Murofushi N, Schmitz G, Theres K, Yamaguchi I (2001) A novel transposon tagging element for obtaining gain-of-function mutants based on a self-stabilizing *Ac* derivative. Plant Mol Biol 45:123–131
49. Ayliffe MA, Pallotta M, Langridge P, Pryor AJ (2007) A barley activation tagging system. Plant Mol Biol 64:329–347
50. Ayliffe MA, Agostino A, Clarke BC, Furbank R, von Caemmerer S, Pryor AJ (2009) Suppression of the barley *uroporphyrinogen III synthase* gene by a *Ds* activation tagging element generates developmental photosensitivity. Plant Cell 21:814–831
51. Mazier M, Botton E, Flamain F, Bouchet JP, Courtial B, Chupeau MC, Chupeau Y, Maisonneuve B, Lucas H (2007) Successful gene tagging in lettuce using the *Tnt1* retrotransposon from tobacco. Plant Physiol 144:18–31
52. Qu S, Desai A, Wing R, Sundaresan V (2008) A versatile transposon-based activation tag vector system for functional genomics in cereals and other monocot plants. Plant Physiol 146:189–199
53. Mora-Garcia S, Vert G, Yin Y, Cano-Delgado A, Cheong H, Chory J (2004) Nuclear protein phosphatases with Kelch-repeat domains modulate the response to brassinosteroids in Arabidopsis. Genes Dev 18:448–460
54. Li J, Lease KA, Tax FE, Walker JC (2001) BRS1, a serine carboxypeptidase, regulates BRI1 signaling in *Arabidopsis thaliana*. Proc Natl Acad Sci USA 98:5916–5921

55. Li J, Wen J, Lease KA, Doke JT, Tax FE, Walker JC (2002) BAK1, an Arabidopsis LRR receptor-like protein kinase, interacts with BRI1 and modulates brassinosteroid signaling. Cell 110:213–222
56. Yuan T, Fujioka S, Takatsuto S, Matsumoto S, Gou X, He K, Russell SD, Li J (2007) *BEN1*, a gene encoding a dihydroflavonol 4-reductase (DFR)-like protein, regulates the levels of brassinosteroids in *Arabidopsis thaliana*. Plant J 51:220–233
57. Zhou A, Wang H, Walker JC, Li J (2004) BRL1, a leucine-rich repeat receptor-like protein kinase, is functionally redundant with BRI1 in regulating Arabidopsis brassinosteroid signaling. Plant J 40:399–409
58. Kang B, Wang H, Nam KH, Li J (2010) Activation-tagged suppressors of a weak brassinosteroid receptor mutant. Mol Plant 3:260–268
59. Guo Z, Fujioka S, Blancaflor EB, Miao S, Gou X, Li J (2010) TCP1 modulates brassinosteroid biosynthesis by regulating the expression of the key biosynthetic gene *DWARF4* in *Arabidopsis thaliana*. Plant Cell 22:1161–1173
60. Becker D (1990) Binary vectors which allow the exchange of plant selectable markers and reporter genes. Nucleic Acids Res 18:203
61. Liu Y-G, Mitsukawa N, Oosumi T, Whittier RF (1995) Efficient isolation and mapping of *Arabidopsis thaliana* T-DNA insert junctions by thermal asymmetric interlaced PCR. Plant J 8:457–463
62. O'Malley RC, Alonso JM, Kim CJ, Leisse TJ, Ecker JR (2007) An adapter ligation-mediated PCR method for high-throughput mapping of T-DNA inserts in the Arabidopsis genome. Nat Protoc 2:2910–2917
63. Clough SJ, Bent AF (1998) Floral dip: a simplified method for *Agrobacterium*-mediated transformation of *Arabidopsis thaliana*. Plant J 16:735–743
64. Altschul SF, Madden TL, Schaffer AA, Zhang J, Zhang Z, Miller W, Lipman DJ (1997) Gapped BLAST and PSI-BLAST: a new generation of protein database search programs. Nucleic Acids Res 25:3389–3402

Chapter 10

Rho GTPase Activity Analysis in Plant Cells

Tongda Xu

Abstract

Rho-family small GTPases are conserved molecular switches of signaling networks in eukaryotic cells, and regulate many cellular responses, such as cytoskeletal reorganization, gene expression, and polarized vesicular trafficking. To understand the functions of Rho GTPase, it is important to investigate how the activity of Rho GTPase is regulated. Plant Rho-like GTPases (ROPs) are known to be regulated by hormones and environmental cues. A rapid activation of ROPs by a stimulus implies a direct signaling role for the ROP GTPases. Here, I describe an ROP activity assay that allows the measurement of ROP GTPase activity that occurs within seconds upon treatment with a stimulus. This method has been successfully used to investigate auxin activation of ROPs in plant cells and will be generally useful for measuring changes in ROP activity with high time resolution (Xu et al., Cell 143:99–110, 2010).

Key words: ROP GTPase, RICs, Auxin, Pull-down assay

1. Introduction

ROP/RAC GTPase (Rho-like GTPases from plants) is a sole plant subfamily of Rho GTPases (1, 2). It functions as a simple binary switch between GDP-bound off mode and GTP-bound on forms. It integrates many upstream signals through RhoGEFs (guanine nucleotide exchange factors), RhoGDIs (guanine nucleotide dissociation inhibitors), and RhoGAPs (GTPase-activating proteins) and regulates multiple downstream effectors, such as RICs (for Rop-interactive CRIB motif-containing proteins) and ICRs (interactor of constitutively active ROPs) (3). Recent studies showed that ROP GTPases modulate cytoskeletal organization in plant cells through their downstream effectors RICs (4–8). RICs contain a CRIB (for Cdc42/Rac-interactive binding) motif that is required for their specific interaction with GTP-bound ROP GTPase (4).

Zhi-Yong Wang and Zhenbiao Yang (eds.), *Plant Signalling Networks: Methods and Protocols*,
Methods in Molecular Biology, vol. 876, DOI 10.1007/978-1-61779-809-2_10,

This specific interaction between RICs and active ROPs provides a basis for the development of methods for measuring ROP activity changes in living plant cells using fluorescence resonance energy transfer (FRET) analysis (6) (Chapter 11, this series) and pull-down assays (9, 10). The FRET-based assay provides information on spatiotemporal changes in ROP activity in living cells as described by Fu (Chapter 11, this series).

The FRET analysis also depends on the specific interaction between RICs and active ROPs. CFP-fused RICs and YFP-fused ROPs are coexpressed in the same cell, and FRET efficiency is used to quantify the amount of interaction between RICs and ROPs, which indicate the relative amount of active ROPs in an individual cell. This method is particularly useful for monitoring the spatio-temporal changes in the activity of an ROP in live cells, but this method is technically challenging and is not as sensitive as the pull-down assay (6, 8). The FRET-based ROP activity assay is described in Chapter 11 in this series.

For pull-down-based ROP activity assay, RICs fused to maltose-binding protein (MBP) is used to "pull down" GTP-bound active Rho GTPase from total protein extracts. Western blotting involving anti-ROP antibody is used to detect the amount of active ROPs pulled down by RICs (9, 11). The *Arabidopsis* genome encodes 11 ROPs that share high sequence similarity. It is difficult to develop specific antibodies for each of the 11 ROPs. To measure the activity of a specific ROP, this ROP is fused with a tag, such as GFP or Myc epitope. Transgenic plants expressing the fusion ROP are used for the isoform-specific ROP activity assay (9). Auxin activation of ROP2 and ROP6 is used as an example to describe this method (9).

Sample preparation is a critical step that affects the variability of pull-down assay. Protoplast provides a powerful system to minimize the variability, especially when we test the multiple samples with different treatments. Since the protoplast samples for different treatments are from the aliquot of the same original protoplast preparation, this greatly reduces the variables among different treatments.

2. Materials

2.1. Plant Materials and Growth Condition

Arabidopsis plants (Col-0) are grown at 22°C on MS agar plates or in soil with 16-h light/8-h dark cycles unless indicated otherwise. 35S::GFP-ROP2 and 35S::GFP-ROP6 transgenic plants are used for pull-down assays. Rosette leaves from 2- to 3-week-old plants are used for protoplast isolation.

2.2. Protoplast Preparation

1. Protoplast enzyme solution (pH 5.6): 1% Cellulase R-10; 0.25% Macerozyme R-10 (*Yakult Honsha, Tokyo, Japan*); 0.4 M mannitol; 10 mM $CaCl_2$; 20 mM Mes-KOH; and 20 mM KCl 0.1% BSA. Cellulase and Macerozyme are added before use.
2. W5 solution (pH 5.6): 154 mM NaCl; 125 mM $CaCl_2$; 5 mM KCl; 5 mM glucose; and 1.5 mM Mes-KOH. This solution is autoclaved before use.

2.3. MBP-RIC1-Conjugated Amylose Beads Purification

1. Amylose beads in 20% ethonal (NEB). Store at 4°C.
2. MBP column binding buffer: 200 mM NaCl; 20 mM Tris–HCl; 1 mM EDTA; 1 mM DTT; 200 μM PMSF (added before use), pH 7.4.

2.4. Pull-Down Assay for Active ROPs

1. Extract Buffer (with or without Triton X): 25 mM Hepes, pH 7.4; 10 mM $MgCl_2$; 100 mM KCl; 5 mM DTT; 5 mM Na_3VO_4; 5 mM NaF; 1 mM PMSF; Proteinase Inhibitor cocktail (Roche); 1% Triton X-100. For extraction buffer without Triton X, TritonX-100 is not excluded. This buffer is prepared before use. To make Na_3VO_4 stock solution, the crystals are dissolved in a tube in a boiling water bath or on a hot plate, and adjust pH to 9–10 using HCl until the color turns bright yellow.
2. Wash buffer: 25 mM Hepes, pH 7.4; 1 mM EDTA; 5 mM $MgCl_2$; 1 mM DTT; 0.5% Triton X-100.

2.5. SDS-PAGE and Western Blotting for Both Active ROPs and Total ROPs

1. Separating buffer (4×): 1.5 M Tris–HCl, pH 8.7.
2. Stacking buffer (4×): 0.5 M Tris–HCl, pH 6.8.
3. 30% Acrylamide/bis solution (39:1) (Biorad) and *N,N,N,N'*-tetramethyl-ethylenediamine (TEMED, Bio-Rad).
4. 10% Ammonium persulfate in water; store at 4°C.
5. Running buffer (5×): 125 mM Tris, 960 mM glycine, 0.5% (w/v) SDS.
6. Prestained molecular weight markers: Kaleidoscope markers (Bio-Rad, CA).
7. Transfer buffer: 25 mM Tris, 190 mM glycine, 20% (v/v) methanol, 0.05% (w/v) SDS.
8. Blocking buffer: 1% fat-free milk (w/v) in phosphate-buffered saline (PBS, pH 7.4).
9. Nitrocellulose membrane from Millipore and 3MM chromatography paper from Whatman.
10. Primary antibody: Rabbit polyclonal IgG Anti-GFP (Santa Cruz Biotech, Inc).
11. Secondary antibody: ECL Anti-Rabbit IgG-HRP (GE Healthcare).
12. Super Signal West Femto Kit (Thermo Scientific) and Bio-Max ML film (Kodak).

3. Methods

Pull-down ROP activity assay is based on the specific interaction between active ROPs and their downstream effectors RICs (4). For example, both RIC1 and RIC4 can specifically bind active-form ROP2 but not inactive-form ROP2 both in vivo and in vitro (Fig. 1) (4, 6). Therefore, we can use MBP-fused RIC1 to pull down active ROP2 from protein extract. This method can be used for the accurate calculation of ROP activity changes in response to the external stimulus, such as chemical treatment or environmental changes (9, 10). Recent work demonstrated that auxin is a key developmental signal that can activate both ROP2 and ROP6 in leaf pavement cells. Thus, we use this as an example to describe pull-down ROP activity assay. GFP-ROP2 or GFP-ROP6 transgenic plants are used to detect the specific ROP2 or ROP6 by GFP antibody (see Note 1).

3.1. Sample Preparation

There are two ways to prepare the samples for pull-down assay: one is to grind plant tissues directly in liquid nitrogen and the other is to take advantage of protoplast. For the first method, the same weight of the plant tissue with different treatments can be grinded for further assay, but this generates greater variablity especially because it is difficult to aliquot the same amount of tissue samples for every treatment. Therefore, we prefer to use protoplasts for sample treatments. Since the protoplast samples for different treatments are aliquoted from the same original sample, the variability will be minimized. We, thus, describe the method involving the use of the protoplast.

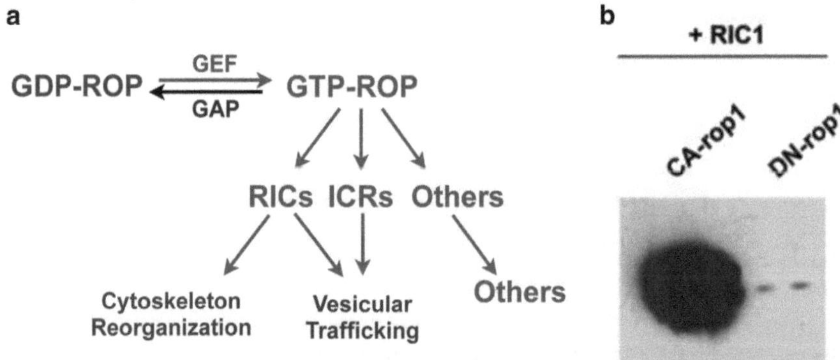

Fig. 1. Rho-like GTPases (ROP) and their downstream effectors in plants. (**a**) A brief scheme about how ROPs' function in plants. ROPs have GDP-bound inactive form and GTP-bound active form. It can be active by guanine nucleotide exchange factors (GEFs) and deactivated by GAPs. Active ROPs can regulate downstream effectors, such as RICs or ICRs, that can further regulate cytoskeleton organization and vesicular trafficking. (**b**) RIC1, as downstream effector of ROPs, can only bind with active-form ROPs. By in vitro pull down, RIC1 can only bind with the constitutive active (CA) ROP1 but not with dominant negative (DN) ROP1 (4).

1. Protoplast isolation method is according to the protocol from J. Sheen's laboratory (9, 12). A 15 ml protoplast enzyme solution is freshly prepared according to Subheading 2.2 and heated at 55°C to inactivate proteases and enhance enzyme solubility (see Note 2). The solution should be clear and brown after passing through a 0.45-μm filter. After cooling down, 20–30 3-week-old true leaves are cut into 0.5–1-mm leaf strips in enzyme solution with fresh razor blades to digest the cell wall. The leaf strips are then incubated in the solution at 23°C for 7–12 h with very gentle shaking.
2. The cells are collected when the leaf strips are dissolved to the solution. The debris is filtered by passing the solution through a miracloth (Cal-biochem). Transfer the pass-through solution to a 50-ml centrifuge tube by blunted 1-ml tips, then add the same volume of W5 solution to cells, and mix gently by inverting the tube. Spin the cells at 200 × *g* for 5 min at room temperature. The green protoplasts are accumulated at the bottom of the tube. The supernatant should be discarded very carefully.
3. The protoplasts are resuspended in around 10 ml W5 solution, and can be stored at 4°C for 1 week (see Note 3).
4. The suspended protoplasts can be evenly aliquoted to a certain amount of samples in 2-ml round-bottom tubes for different treatments (see Note 4). Here, we isolated the protoplasts from GFP-ROP2 and GFP-ROP6 transgenic plants to test the ROP2 activity and ROP6 activity after auxin treatment. The even aliquots of samples are treated with different concentrations of auxin (from 1 nM to 10 μM) for the same period (2 min) or with the same concentration of auxin (100 nM) for different period (from 30 s to 8 min) (9). The samples are precipitated and frozen in liquid nitrogen rapidly after treatment (see Note 5). Frozen samples can be stored in −80°C freezer before further analysis.

3.2. Preparation of MBP-RIC1-Conjugated Amylose Beads

1. For the preparation of MBP-RIC1 beads, RIC1 was fused with MBP in pMAL21 construct as described previously and transformed to BL21 strain (4). Fusion protein was expressed at 28°C for 4 h after induction with 1 mM isopropyl-β-D-thiogalactopyranoside (IPTG). Cell cultures were centrifuged at 4,000 × *g* for 10 min.
2. Cell pellets were resuspended in 20 ml column buffer (Subheading 2.3) and sonicated by using 15-s pulses for eight times with a 10-s interval. The supernatant was obtained by centrifugation at 12,000 × *g* for 30 min at 4°C. The supernatant was transferred to another tube for binding assay.
3. Approximately 20 ml supernatant was mixed with 1–2 ml amylose beads (Biolabs), which were prewashed by column buffer

(three times, 10 min each). The beads are incubated in cold room for 2 h with gentle shaking.

4. The beads were centrifuged and washed with the column buffer three times, 10 min each time. These MBP-RIC1-conjugated beads are ready for use or can be stored in −80°C freezer for future use.

3.3. Pull-Down Procedure

1. Before pull-down procedure, all the buffers or other materials need to be cooled down in a refrigerator (see Note 1). The protoplast samples from Subheading 3.1 are thawed on ice, and then add 1 ml cold extraction buffer with Triton X that is described in Subheading 2.4. The samples are mixed well and then kept in 4°C cold room with gentle shaking for 30 min. This step helps to dissolve the protoplast and break the cells (see Note 6).
2. Well-dissolved samples are centrifuged at 10,000 × *g* for 10 min in cold room to precipitate the debris. The supernatant is transferred to a new tube and mixed with the same volume of extraction buffer without Triton X.
3. To avoid the nonspecific binding by amylose beads, we first add 50 μl blank amylose beads (prewashed by extraction buffer three times) to the samples, incubate for 15 min in cold room, and then centrifuge the beads at 200 × *g* for 1 min.
4. From this step, we take and save 50 μl supernatant for detecting the total ROP level (GTP and GDP) from each sample. We use Western blotting to detect the amount of total ROP2 or ROP6 in each sample that is described in Subheading 3.4.
5. The rest of the supernatant is mixed with 100 μl MBP-RIC1 beads and incubated in 4°C cold room for 2 h with gentle shaking. Make sure that the beads are always suspended during incubation.
6. The beads are pelleted by centrifuge at 200 × *g* for 2 min in cold room, and then washed by 1 ml washing buffer three times (5 min each time). During washing, we prefer to leave some washing buffer in the tube to avoid the loss of beads (see Note 7).
7. We leave 20 μl washing buffer with the beads in the last washing, add 5 μl 5× sampling buffer to the beads, and incubate the tube in the boiled water bath for at least 3 min.
8. The beads are then centrifuged at 3,000 × *g* for 1 min, and the supernatant is loaded to SDS-PAGE gel to separate the proteins for Western blotting.

3.4. SDS-PAGE and Western Blotting

1. Prepare two 1.5-mm-thick 10% separating gels by mixing 2.5 ml of 4× separating buffer, 3.3 ml acrylamide/bis solution (30%), 50 μl 20% SDS solution, 4.1 ml water, 50 μl 10%

ammonium persulfate solution, and 10 μl TEMED. The gel is covered by 2 ml water at the top after poured into the vertical gel tank (Bio-Rad 165-8000). The gel is then incubated at 37°C for 30 min. Discard the water after the gel is polymerized.

2. Prepare the stacking gel by mixing 1.25 ml of 4× stacking buffer, 0.67 ml acrylamide/bis solution (30%), 25 μl 20% SDS solution, 3.075 ml water, 25 μl 10% ammonium persulfate solution, and 5 μl TEMED. Pour the stacking gel onto the top of separating gel, add the 10-well comb, and incubate the gel at 37°C for 30–60 min.
3. The comb is then removed carefully after the gel is solidified. After setting up the electrophoresis system, the running buffer (precooled) is added to the upper chamber that can cover the wells of the gel, and 25 μl of the sample is loaded to each well by pipettes. The prestained protein ladder is loaded to the first well. One gel is loaded with the samples eluted from pull-down beads from Subheading 3.3, step 8. One gel is loaded with the total proteins from Subheading 3.3, step 4. Two gels are running totally parallel to each other.
4. Add the running buffer to the lower chambers of the gel unit and run the gel at 75 V for 30 min until the proteins run into separating gel, and then continue running the gel at 125 V for 1 h until the dye front runs out of the gel. The whole procedure is done in 4°C cold room to minimize protein degradation.
5. Once the gel running is finished, the gel can be soaked in transfer buffer before transferring to the membrane. Nitrocellulose membrane (presoaked in transfer buffer) and the gel are stacked together and loaded to electric transfer cassette (Bio-Rad) with two layers of 3-mm filter papers and two layers of forms at each side. The gel side should be close to the negative anode of the power supply. Avoid the bubbles in the cassette.
6. The membranes' transfer is done in cold room at 12 V overnight or at 120 V for 1.5 h.
7. The membranes are then washed briefly with 1× PBS buffer and soaked in blocking buffer for at least 1 h at room temperature or overnight in cold room with gentle shaking (see Note 8).
8. The GFP antibody solution (1:500 diluted in blocking buffer) (Santa Cruz Biotech) is then added to the membrane, and then incubated for 3 h at room temperature or overnight in cold room.
9. Wash the membranes with 1× PBS three times, 10 min each time, and then incubate the membrane in the secondary antibody solution (1:1,000 diluted in 1× PBS). Then, incubate at room temperature for 2–3 h.
10. The membranes are washed with 1× PBS three times, 10 min each time. These membranes are ready for signal detection or can be stored at −20°C before use.

3.5. Data Collection and Analysis

1. The membranes from the previous step are soaked in the reaction buffer from Super Signal West Femto Kit (Thermo Scientific) for a few minutes and exposed to Bio-Max ML film (Kodak) in darkroom. The time of exposure can range from a few seconds to a few minutes depending on the signal intensity. The membranes for active ROPs and total ROPs are treated equally to each other and exposed to the same film.
2. The signal intensity for each sample, which indicates the amount of ROP protein, is measured by Metamorph software. The intensity of both GTP-bound and GDP-bound ROP2 or ROP6 multiplied by 40 (2 ml/50 μl) is considered as the total ROP2 or ROP6 amount from the original samples. The intensity of GTP-bound ROP2 multiplied by 40/39 [2 ml/(2 ml to 50 μl)] is considered as the active ROP2 or ROP6 amount from the original samples.
3. The relative active ROP2 or ROP6 levels (GTP bound) are normalized to the total amount of ROP2 or ROP6 (both GDP and GTP bound). We divide the amount of active ROPs by the amount of total ROPs to obtain the relative amount of active ROPs in each sample.
4. The level of GTP-bound ROPs relative to the control (0 nM NAA at 0 min) was calculated by dividing the amount of normalized GTP-bound ROP2 or ROP6 from each treatment by the normalized amount from the control, which is defined as "1" (Fig. 2) (9).

4. Notes

1. The whole procedure of the pull-down assay needs to be done in cold room to keep everything cold during the procedure. This minimizes the degradation of the proteins and improves the accuracy of the method.
2. The buffers for protoplasts should be carefully prepared to assure that the concentration of every component is accurate because the protoplast is fragile and a minor change in the osmotic pressure may produce unhealthy or damaged protoplasts. Use the light microscope to check whether the protoplast is healthy or not. If the protoplast is prepared well, we will see all round-shaped protoplast but rarely see broken ones.
3. It is better to use fresh protoplast for different treatment. If you use the old protoplast from 4°C refrigerator, you have to keep the samples at room temperature for a while to make the protoplasts most active for different treatments.
4. To aliquot the protoplast for each treatment, you should continuously shake the tube in order to keep the protoplasts

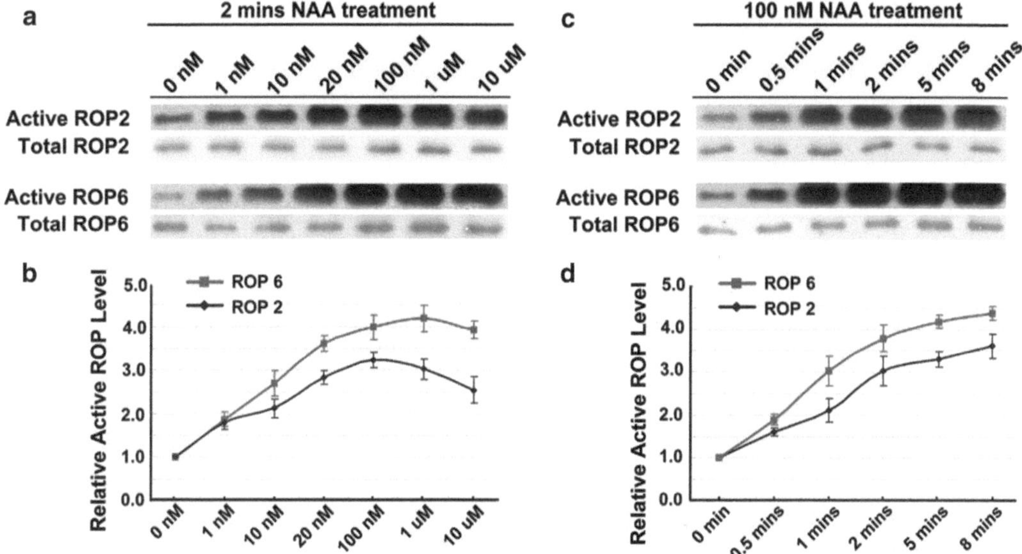

Fig. 2. Pull-down assay for auxin activation of ROP2 and ROP6 in plant cells. (**a**, **b**) Pull-down assay of active ROP2 and ROP6 in response to auxin. Protoplasts from leaves of transgenic GFP-ROP2 or -ROP6 seedlings were treated with the indicated concentrations of NAA for 2 min (**a**) or treated with 100 nM NAA for the indicated times (**b**). GTP-bound active GFP-ROP2 or -ROP6 and total GFP-ROP2 or -ROP6 were detected by Western blotting. ROP2 and ROP6 experiments were conducted in parallel under identical conditions (9). (**c**, **d**) Quantitative analysis of data from (**a**) and (**b**). The relative ROP2 or ROP6 activity level was determined as the amount of GTP-bound ROP2 or ROP6 divided by the amount of total GFP-ROP2 or -ROP6. The relative ROP activity in different treatments was standardized to that from mock-treated control, which was arbitrarily defined as "1".

evenly suspended; this assures that each aliquot contains the same number of protoplasts.

5. During treatments, add auxin to the wall of the tube, mix the sample with auxin, and start to time immediately. When time is up, spin the sample for 5 s, pour the liquid out very fast, and put it in the liquid nitrogen. It is important that these steps are executed in an identical manner in different treatments to minimize the experimental variability.
6. Before pull down, the protoplast samples should be dissolved well in extraction buffer. The liquid should be green and clear. Otherwise, some samples will be lost in the next step.
7. The step for the washing of the beads after the pull down is very critical because the beads can be easily lost during the washing (since the beads can be discarded by mistake when discarding the supernatant). So it is a good practice to leave some washing buffer with the beads to avoid touching the beads at the bottom of the tube.
8. In Western blotting, it is better to block the membrane by blocking buffer overnight at 4°C cold room. It can help to improve the quality of the image in the final step.

References

1. Yang Z (2002) Small GTPases: versatile signaling switches in plants. Plant Cell 14(Suppl): S375–388
2. Gu Y, Wang Z, Yang Z (2004) ROP/RAC GTPase: an old new master regulator for plant signaling. Curr Opin Plant Biol 7:527–536
3. Nagawa S, Xu T, Yang Z (2010) RHO GTPase in plants: conservation and invention of regulators and effectors. Small GTPases 1:78–88
4. Wu G, Gu Y, Li S, Yang Z (2001) A genome-wide analysis of Arabidopsis Rop-interactive CRIB motif-containing proteins that act as Rop GTPase targets. Plant Cell 13:2841–2856
5. Fu Y, Li H, Yang Z (2002) The ROP2 GTPase controls the formation of cortical fine F-actin and the early phase of directional cell expansion during Arabidopsis organogenesis. Plant Cell 14:777–794
6. 'Fu Y, Gu Y, Zheng Z, Wasteneys G, Yang Z (2005) Arabidopsis interdigitating cell growth requires two antagonistic pathways with opposing action on cell morphogenesis. Cell 120:687–700
7. Gu Y, Fu Y, Dowd P, Li S, Vernoud V, Gilroy S, Yang Z (2005) A Rho family GTPase controls actin dynamics and tip growth via two counteracting downstream pathways in pollen tubes. J Cell Biol 169:127–138
8. Fu Y, Xu T, Zhu L, Wen M, Yang Z (2009) A ROP GTPase signaling pathway controls cortical microtubule ordering and cell expansion in Arabidopsis. Curr Biol 19:1827–1832
9. Xu T, Wen M, Nagawa S, Fu Y, Chen JG, Wu MJ, Perrot-Rechenmann C, Friml J, Jones AM, Yang Z (2010) Cell surface- and rho GTPase-based auxin signaling controls cellular interdigitation in Arabidopsis. Cell 143:99–110
10. Baxter-Burrell A, Yang Z, Springer PS, Bailey-Serres J (2002) RopGAP4-dependent Rop GTPase rheostat control of Arabidopsis oxygen deprivation tolerance. Science 296:2026–2028
11. Tao LZ, Cheung AY, Wu HM (2002) Plant Rac-like GTPases are activated by auxin and mediate auxin-responsive gene expression. Plant Cell 14:2745–2760
12. Sheen J (2001) Signal transduction in maize and Arabidopsis mesophyll protoplasts. Plant Physiol 127:1466–1475

Chapter 11

Analysis of In Vivo ROP GTPase Activity at the Subcellular Level by Fluorescence Resonance Energy Transfer Microscopy

Lei Zhu and Ying Fu

Abstract

Proteins generally interact with some other proteins to achieve their cellular functions. Fluorescence resonance energy transfer (FRET) microscopy provides a powerful technique to elucidate such interactions in vivo. FRET occurs when two properly chosen fluorophores are sufficiently close (less than 10 nm). Aided by multiple colored fluorescent proteins (FPs), FRET microscopy has been widely used in live cells for detection of protein–protein interaction and in some cases protein activity in a real-time in vivo manner, which contributes to the understanding of the mechanisms for the regulation of many cellular activities, such as signal transduction pathways. Here, we describe a convenient and fast FRET imaging microscopy involving transiently expressed proteins fused with an FRET pair of fluorescent proteins (e.g., cyn fluorescent protein and yellow fluorescent protein). We describe an example of the FRET-based assay used to analyze ROP GTPase activity in live plant cells.

Key words: Fluorescence resonance energy transfer, Protein–protein interaction, Cyn fluorescent protein, Yellow fluorescent protein, Signal transduction, Transient expression

1. Introduction

The spatiotemporal regulation of protein–protein interactions is a pivotal mechanism for the control of many cellular processes, including signal transduction in response to intra- and intercellular signals. Therefore, the elucidation of such interactions in time and space within a specific cell will provide insights into the biological mechanisms underscoring these important cellular processes and their regulation (1–3). Several approaches have been broadly used to detect protein–protein interactions, although each has its advantages and technical limitations. Traditional in vitro pull-down methods allow the detection of direct interactions but may not

Zhi-Yong Wang and Zhenbiao Yang (eds.), *Plant Signalling Networks: Methods and Protocols*, Methods in Molecular Biology, vol. 876, DOI 10.1007/978-1-61779-809-2_11,

reflect in vivo interactions between proteins in the native states within living cells, especially those that occur transiently. Coimmunoprecipitation enables quantifying protein interactions that occur in the native state (2, 4, 5), but is unable to determine whether the interactions are direct and to visualize the spatiotemporal dynamics of the interactions that occurs at the cellular or subcellular level.

Fluorescence resonance energy transfer (FRET) imaging microscopy can overcome the above limitations. FRET occurs when two properly chosen fluorophores (termed as donor and acceptor) are sufficiently close (less than 10 nm). When the donor molecule absorbs a quantum of light, an electron is boosted up into an excited state with a higher energy. When this electron relaxes back to a nonexcited state, the energy is transferred to the acceptor (6, 7). To be a FRET pair, the donor and the acceptor must display overlapping emission/absorption spectra for efficient energy transfer, and the reasonable separation in emission spectra between donor and acceptor to minimize the cross talk (7–10). Over the past decade, the FRET imaging microscopy becomes a powerful technique for studying protein–protein interactions inside living cells with higher sensitivity as well as the improved spatial and temporal resolution. This is due to the improving fluorescence microscopy, aided by the development of new fluorescent probes, especially the introduction of the multiple colored fluorescent proteins (FPs) to FRET-based imaging (7, 8, 10–12).

Rho-family small guanine nucleotide-binding proteins (GTPases) are key molecular switches of a wide range of signaling pathways that regulate fundamental cellular processes in different eukaryotic organisms (13–15). A common mechanism for a small GTPase taking action is the interaction of the activated small GTPase with multiple downstream effectors through its effector-binding domain (14, 16, 17). Therefore, studies on the interaction between small GTPases and different effectors will contribute to the discovery of the function and the regulatory mechanisms of different Rho-signaling pathways. Furthermore, an important hallmark of the interactions between small GTPases and their effectors is that the interactions are specific to the activated form of GTPases (16, 18, 19). Thus, the FRET-based assay of the interactions between Rho GTPases and their effectors has been widely used to monitor the spatiotemporal dynamics of Rho GTPase activities within live cells (3, 20–22).

Plants possess a unique subfamily of Rho-GTPases, termed ROP (*Rho*-related GTPase from *p*lants) (13–18, 23). As the sole subfamily of signaling small GTPases in plants, ROP GTPases become critical signal integrators and coordinators to regulate plant development and responses to hormones, pathogens, and abiotic stresses through interacting with various effectors (13–17, 23). Here, we use a member of ROP family GTPases, ROP6, and its downstream effector, RIC1 (*ROP-i*nteractive *C*RIB motif containing protein *1*), as an example for FRET-based analysis of ROP6 activation. ROP6 and RIC1 are fused to a FRET pair of fluorescent proteins, yellow fluorescent protein (YFP) and cyn fluorescent protein (CFP), respectively. A particle-bombardment method is used to introduce CFP-RIC1 and YFP-ROP6 constructs into leaf pavement cells for transient expression, which provides an efficient method that allows rapid analysis of the interaction within living cells. The FRET imaging microscopy and the analysis of FRET efficiency are used to illustrate the dynamic interaction between ROP6 and RIC1 in a quantitative manner at the subcellular resolution (22, 24).

2. Materials

2.1. Plasmid Preparations

pBS35S:YFP was kindly provided by Dr. A. von Amin. *ROP6* was amplified by PCR and cloned into the Bgl II and BamH I sites of *pBS35S:YFP*. *DN-rop6* (D121A, dominant negative mutant of ROP6) was generated by use of the Quick Change Site-Directed Mutagenesis Kit (Qiagen) and cloned into Bgl II and BamH I sites of *pBS35S:YFP*. *CFP-RIC1* was constructed in the *pBI221* vector (Clontech) using BamH I and Sst I sites (21, 24, 25).

All plasmids were amplified in *Escherichia coli* strain Top 10 and purified using plasmid midi or mini kits according to the manufacturer's instructions (Qiagen).

2.2. Plant Materials

Grow *Arabidopsis thaliana* ecotypes Col-0 at 22°C in soil with 16-h light/8-h dark cycles. Collect expanding rosette leaves of 0.8–1.2 cm in length from 3- to 6-week-old plants (21, 24, 25).

2.3. Particle-Bombardment Supplies

A PDS-1000/He particle delivery system (Bio-Rad Laboratories) is used.

1. Supplies for particle bombardment are purchased from Bio-Rad Laboratories, including 1.0-μm microcarrier (gold particles), macrocarrier, stopping screen, and 650-psi rupture disks.
2. Additional regents include sterile water, 70% ethanol, 100% ethanol, 50% glycerol, 2.5 M $CaCl_2$, and 0.1 M spermidine.

3. Methods

3.1. Particle Bombardment

3.1.1. Wash Microcarriers

1. Weigh out 30 mg of 1.0-μm gold particles into a 1.5-ml microcentrifuge tube.
2. Add 1 ml of 70% ethanol (*v*/*v*) and vortex vigorously for 3–5 min.
3. Allow the particles to soak in 70% ethanol for 15 min.

4. Spin for 5 s, and then discard the supernatant.
5. Add 1 ml of sterile water and vortex vigorously for 1 min.
6. Spin for 5 s, and then discard the supernatant.
7. Repeat steps 5 and 6 two times.
8. Add 500 μl sterile 50% glycerol and store the washed gold at 4°C.

3.1.2. Coat Washed Microcarriers with DNA

The following procedure is for one bombardment shot; if more bombardments are needed, adjust the quantities accordingly. The concentration of purified plasmid DNA is adjusted to 0.5 μg/μl.

1. Vortex the stock microcarriers in 50% glycerol vigorously for 5 min to resuspend the gold particles.
2. Remove 8–10 μl of microcarriers to a 1.5-ml microcentrifuge tube, and then add the following in order: DNA (2 μl of *pBI221:CFP-RIC1* and 1 μl of *pBS35S:YFP-ROP6* or *pBS35SYFP-DN-rop6*), 30 μl of 2.5 M $CaCl_2$, and 12 μl of 0.1 M spermidine (see Note 1).
3. Vortex vigorously for 15 min, and then allow the coated microcarriers to settle for 2 min.
4. Spin for 2 s, and then discard the supernatant.
5. Add 70 μl of 70% ethanol, spin for 2 s, and then discard the supernatant.
6. Add 70 μl of 100% ethanol, spin for 2 s, and then discard the supernatant.
7. Add 8–10 μl of 100% ethanol. Gently tap the tube several times to resuspend the pellet and then vortex at low speed for 1 min.

3.1.3. Perform A Bombardment

1. Open the helium cylinder valve and adjust the pressure to 850 psi (that is 200 psi in excess of desired rupture disk).
2. Power ON.
3. Load rupture disk into retaining cap and secure the retaining cap to the end of gas acceleration tube and tighten.
4. Apply the DNA-coated gold particles on the macrocarrier, and spread the particles into a circle (1 cm in diameter) in the middle of the macrocarrier.
5. Load macrocarrier into the holder and turn it upside down. Put the holder and the stopping screen into microcarrier launch assembly.
6. Place the microcarrier launch assembly in chamber, place the target leaves in a distance about 8 or 11 cm below, and then close the door.
7. Evacuate the chamber, and hold vacuum at 27″ of Hg.

8. Press and hold the Fire button until rupture disk bursts and helium pressure gauge drops to zero.
9. Release fire button.
10. Release vacuum from chamber.
11. Remove leaves from chamber and place them in water.
12. Unload macrocarrier and stopping screen from microcarrier launch assembly (see Note 2); unload spent rupture disk.
13. Repeat steps 3–12 if more shots are needed.
14. Shut down the whole system according to the manufacturer's instruction.

3.2. Confocal Microscopy for FRET Analysis

Transformed cells are observed 4–6 h after particle bombardment using the Leica TCS SP2 confocal microscope, which allows flexible selection of emission bandwidths to minimize bleach through (see Note 3). The 15% power of 442-nm laser or 514-nm laser is set for excitation. CFP signal is collected using 465–490-nm bandwidths and YFP/FRET signal is collected using 560–640-nm bandwidths (21, 24). For 3D reconstruction, optical sections are scanned and captured at 1.0-μm increment using the Leica software.

1. Perform control experiments with donor (*pBI221:CFP-RIC1*) alone and acceptor (*pBS35S:YFP-ROP6*) alone. Collect images of (1) donor and acceptor emission with 442-nm laser excitation and (2) acceptor emission with 514-nm laser excitation.
2. Perform FRET experiments with both donor and acceptor. Collect signals of (1) donor and FRET emission with 442-nm laser excitation and (2) acceptor emission with 514-nm laser excitation. An example of FRET images is shown in Fig. 1.

3.3. Calculation of FRET Efficiency

1. In the dark area outside the cell in each image, draw a region of interest (ROI) and use the average fluorescence intensity of the ROI as background noise to process background subtraction function of MetaMorph software.
2. Draw an ROI around the fluorescent PM and cortex of the cell in the image collected from CFP channel, and then copy the ROI to YFP image as well as FRET image at exactly the same place using MetaMorph software (see Note 4) (24). Measure the average fluorescence intensity of the ROI in each image. An Illustration of how to draw an ROI is shown in Fig. 1.
3. Calculate correction factors from CFP donor-alone and YFP acceptor-alone controls:
   (a) CFP channel emission of YFP acceptor (excited by 442-nm laser) divided by YFP channel emission of YFP acceptor (excited by 514-nm laser)

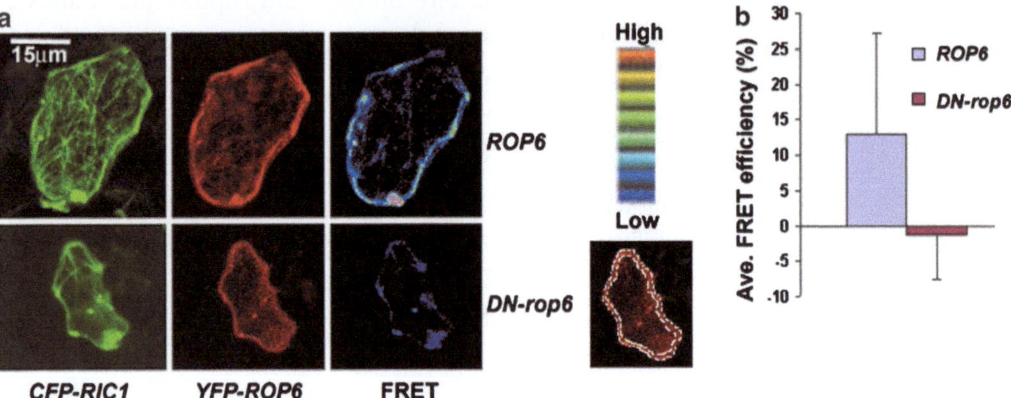

Fig. 1. ROP6 physically interacts with RIC1 in vivo. (**a**) Fluorescence resonance energy transfer (FRET) analysis was performed in pavement cells expressing CFP-RIC1 (*pseudocolored green*) and YFP-ROP6 or YFP-DN-rop6 (*pseudocolored red*). The pseudocolor scale was used to indicate the FRET signal intensity. The image under the pseudocolor scale indicates the region of the cell where the fluorescence intensity was measured. (**b**) Quantitative analysis of FRET efficiency. All data are mean ± standard deviation. Reproduced from ref. 24 with the permission from Elsevier Science.

(b) YFP channel emission of CFP donor (excited by 442-nm laser) divided by CFP channel emission of CFP donor (excited by 442-nm laser)

(c) YFP channel emission of YFP (excited by 442-nm laser) divided by YFP channel emission of YFP (excited by 514-nm laser)

4. Collect data from FRET experiments (cells coexpressing CFP donor and YFP acceptor). Data are as follows:

   (A) CFP channel emission with 442-nm laser excitation

   (B) YFP channel emission with 442-nm laser excitation

   (C) YFP channel emission with 514-nm laser excitation

5. Correct FRET signal: Corrected FRET signal $= B - b \times A - (c - a \times b) \times C$.

6. Calculate FRET efficiency: FRET efficiency = corrected FRET signal/C (see Note 5). An example result is shown in Fig. 1.

4. Notes

1. The amount of $CaCl_2$ and spermidine for DNA coating is decided according to the total volume of DNA that is added to microcarriers in 50% glycerol; the ratio of DNA:$CaCl_2$ is 1:10 (v/v) and the ratio of DNA:spermidine is 1:4 (v/v). Therefore, in control experiments with donor alone or acceptor alone, the amount of $CaCl_2$ and spermidine for DNA coating needs to be adjusted accordingly.

2. The stop screens can be reused. Recycled stop screens are cleaned and either autoclaved or stored in 70% ethanol.
3. Other equivalent confocal microscopes made by different manufacturers can be used as well.
4. Copy ROI function of MetaMorph ensures that the measurements of average fluorescence intensity are done precisely within the same region.
5. FRET efficiency represents the percent of emission in YFP channel that results from FRET. The corrected FRET signal is normalized with the acceptor amount for comparison. This is to exclude the interference of cell-to-cell variability in transient expression levels.

Acknowledgments

We would like to thank Dr. A. von Amin for kindly providing *pBS35S:YFP*, and Dr. Jaeung Hwang for assistance in calculation of FRET efficiency. This work was supported by National Natural Science Foundation of China (grant #30970173 to L.Z. and grant # 90817105 to Y.F.).

References

1. Stein A, Pache RA, Bernado P, Pons M, Aloy P (2009) Dynamic interactions of proteins in complex networks: a more structured view. FEBS J 276:5390–5405
2. Monti M, Orru S, Pagnozzi D, Pucci P (2005) Interaction proteomics. Biosci Rep 25:45–56
3. Nakamura T, Aoki K, Matsuda M (2005) Monitoring spatio-temporal regulation of Ras and Rho GTPase with GFP-based FRET probes. Methods 37:146–153
4. Miernyk JA, Thelen JJ (2008) Biochemical approaches for discovering protein-protein interactions. Plant J 53:597–609
5. Ren L, Emery D, Kaboord B, Chang E, Qoronfleh MW (2003) Improved immunomatrix methods to detect protein:protein interactions. J Biochem Biophys Methods 57:143–157
6. Wu P, Brand L (1994) Resonance energy transfer: methods and applications. Anal Biochem 218:1–13
7. Clegg RM (2002) FRET tells us about proximities, distances, orientations and dynamic properties. J Biotechnol 82:177–179
8. Truong K, Ikura M (2001) The use of FRET imaging microscopy to detect protein-protein interactions and protein conformational changes in vivo. Curr Opin Struct Biol 11:573–578
9. Pollok BA, Heim R (1999) Using GFP in FRET-based applications. Trends Cell Biol 9:57–60
10. Kenworthy AK (2001) Imaging protein–protein interactions using fluorescence resonance energy transfer microscopy. Methods 24:289–296
11. Ai HW, Hazelwood KL, Davidson MW, Campbell RE (2008) Fluorescent protein FRET pairs for ratiometric imaging of dual biosensors. Nat Methods 5:401–403
12. Piston DW, Kremers GJ (2007) Fluorescent protein FRET: the good, the bad and the ugly. Trends Biochem Sci 32:407–414
13. Brembu T, Winge P, Bones AM, Yang Z (2006) A RHOse by any other name: a comparative analysis of animal and plant Rho GTPases. Cell Res 16:435–445

14. Fu Y, Kawasaki T, Shimamoto K, Yang Z (2006) ROP/RAC GTPases. In: Yang Z (ed) Intracellular signaling in plants. Wiley-Blackwell, OXford, pp 64–99
15. Yang Z, Fu Y (2007) ROP/RAC GTPase signaling. Curr Opin Plant Biol 10:490–494
16. Yang Z (2002) Small GTPases: versatile signaling switches in plants. Plant Cell 14(Suppl): S375–S388
17. Zheng ZL, Yang Z (2000) The Rop GTPase: an emerging signaling switch in plants. Plant Mol Biol 44:1–9
18. Berken A, Wittinghofer A (2008) Structure and function of Rho-type molecular switches in plants. Plant Physiol Biochem 46:380–393
19. Wu HM, Hazak O, Cheung AY, Yalovsky S (2011) RAC/ROP GTPases and auxin signaling. Plant Cell 23:1208–1218
20. Sekar RB, Periasamy A (2003) Fluorescence resonance energy transfer (FRET) microscopy imaging of live cell protein localizations. J Cell Biol 160:629–633
21. Fu Y, Gu Y, Zheng Z, Wasteneys G, Yang Z (2005) Arabidopsis interdigitating cell growth requires two antagonistic pathways with opposing action on cell morphogenesis. Cell 120:687–700
22. Hwang JU, Gu Y, Lee YJ, Yang Z (2005) Oscillatory ROP GTPase activation leads the oscillatory polarized growth of pollen tubes. Mol Biol Cell 16:5385–5399
23. Berken A (2006) ROPs in the spotlight of plant signal transduction. Cell Mol Life Sci 63:2446–2459
24. Fu Y, Xu T, Zhu L, Wen M, Yang Z (2009) A ROP GTPase signaling pathway controls cortical microtubule ordering and cell expansion in Arabidopsis. Curr Biol 19:1827–1832
25. Fu Y, Li H, Yang Z (2002) The ROP2 GTPase controls the formation of cortical fine F-actin and the early phase of directional cell expansion during Arabidopsis organogenesis. Plant Cell 14:777–794

Chapter 12

In Vivo Ubiquitination Assay by Agroinfiltration

Lijing Liu, Qingzhen Zhao, and Qi Xie

Abstract

The ubiquitination/proteasome system is involved in nearly all plant signaling processes. Many signaling components are degraded by the 26S proteasome upon ubiquitination by specific E3 ubiquitin ligase. However, due to technical limitations, only a few pairs of E3 ligase–substrate interactions have been directly demonstrated in plants. The method described here provides an efficient way to detect E3-mediated protein ubiquitination in vivo by agroinfiltration in *Nicotiana benthamiana*. This assay allows for fast and reliable detection of the specific interaction between the substrate and the E3 ligase, the effect of E3 ligase on substrate ubiquitination and degradation, and the effects of proteasome inhibitor, such as MG132, on substrate stability.

Key words: Ubiquitination assay, Agroinfiltration, E3 ligase, Substrate, In vivo

1. Introduction

The ubiquitin/26S proteasome system (UPS) is a key mechanism in eukaryotic cells to regulate the degradation of specific proteins. Protein degradation by the UPS pathway involves multi-protein complexes, including the E1 ubiquitin-activating enzyme, E2 ubiquitin-conjugating enzyme, E3 ubiquitin protein ligase, 26S proteasome, and ubiquitin (1). In the Arabidopsis genome, there are 2 E1s, 37 E2s (plus 8 E2 variants), and more than 1,300 E3s (2). UPS plays important roles in a wide range of processes, such as plant stress response and development signaling (3, 4). It is known that E3s provide specificity for recognizing substrates, and the E3 ligase–substrate interactions are key mechanisms for regulating protein degradation. In mammals, the interaction and specificity between E3s and substrates can be easily detected by transient expression systems in cultured cell lines (5–7). However, in plants, such a convenient assay has not yet been established. Although some assays based on protein transient expression in protoplast

Zhi-Yong Wang and Zhenbiao Yang (eds.), *Plant Signalling Networks: Methods and Protocols*, Methods in Molecular Biology, vol. 876, DOI 10.1007/978-1-61779-809-2_12,

have been successful, the technical difficulty and low expression level limited their application (8). To fill this gap, we have developed an efficient ubiquitination assay using transient transformation by Agrobacterium infiltration (agroinfiltration) of *Nicotiana benthamiana* leaves (9). The binary vectors for expression of the putative E3 and substrate are transformed into *Agrobacterium* strain EHA105 separately. Then, the Agrobacteria with different combinations are coinfiltrated into *N. benthamiana* with another agrobacteria containing p19 expression cassette, a suppressor of gene silencing from tomato bushy stunt virus (10). It has been shown that p19 can enhance the protein expression up to 50–100 times by stabilizing RNA transcripts (11). After samples are collected, the effect of E3 ligase on substrate ubiquitination and the effect of MG132 on substrate degradation are detected by Western blot, and the interaction between the putative substrate and E3 ligase can be analyzed by immunoprecipitation.

2. Materials

2.1. Nicotiana benthamiana and Agroinfiltration

1. Wild-type *N. benthamiana* seeds.
2. Soil, vermiculite, 9 cm × 9 cm pots and a 22°C growth chamber.
3. *Agrobacterium tumefaciens* strain EHA105 transformed with binary vectors for overexpressing proteins to be tested.
4. LB medium: 10 g/L trytone, 5 g/L yeast extract, and 10 g/L NaCl.
5. 2-(*N*-morpholine)-ethanesulfonic acid (MES): 1 M stock. Dissolve it in H_2O, adjust the pH to 5.6 with NaOH, and store it at room temperature.
6. Acetosyringone (AS): 100 mM stock in DMSO and store it at −20°C.
7. 10 mM $MgCl_2$ solution: Dissolve in H_2O and store it at room temperature.
8. 1-mL Disposable syringe.

2.2. Protein Extraction

1. Native buffer: 50 mM Tris–MES, pH 8.0, 0.5 M sucrose, 1 mM $MgCl_2$, 10 mM EDTA, 5 mM DTT, and protease inhibitor cocktail CompleteMini tablets (Roche; add before use).
2. Denaturing buffer: 50 mM Tris–HCl, pH 7.5, 150 mM NaCl, 0.1% NP-40, 4 M urea, and 1 mM phenylmethylsulfonyl fluoride (PMSF).

2.3. Immunoprecipitation

1. MG132: 10 mM stock, MG132 powder is dissolved into DMSO. The solution can be stored at −80°C.

2. A shaker: Any instrument that can shake or invert the tubes.
3. A cold room at 4°C.
4. Protein G agarose beads (Roche).

2.4. SDS-Polyacrylamide Gel Electrophoresis

1. 30% Acrylamide (Acr)-Bis: Put 30 g of acrylamide and 0.8 g of Bis to 100 mL H_2O and filter through 0.2-μm filter. Store it at 4°C.
2. 10% Ammonium persulfate (APS): 0.1 g of APS in 1 mL H_2O, Store it at 4°C.
3. 10% SDS: 10 g of SDS dissolved in 100 mL H_2O. Store it at room temperature.
4. 1 M Tris–HCl, pH 6.8: Dissolve Tris in H_2O, and adjust the solution pH to 6.8 with HCl. After autoclaved, it can be stored at room temperature.
5. 1.5 M Tris–HCl, pH 8.8: Dissolve Tris in H_2O, and adjust the solution pH to 8.8 with HCl. After autoclaved, it can be stored at room temperature.
6. TEMED (from AMRESCO).
7. 5 mL Separating gel: 1.9 mL of H_2O, 1.7 mL of 30% Acr-Bis, 1.3 mL of 1.5 M Tris–HCl, pH 8.8, 0.05 mL of 10% SDS, 0.05 of mL 10% APS, 2 μL of TEMED.
8. 2 mL Stacking gel: 1.4 mL of H_2O, 0.33 mL of 30% Acr-Bis, 0.33 mL of 1 M Tris–HCl, pH 6.8, 0.02 mL of 10% SDS, 0.02 mL of 10% APS, 2 μL of TEMED.

2.5. Western Blot

1. Running buffer: 3.03 g/L Tris base, 14.4 g/L glycine, and 1 g/L SDS.
2. 4× SDS sample buffer: 0.25 M Tris, pH 6.8, 8% SDS, 40% glycerol, 0.005% bromophenol blue, and 20% 2-mercaptoethanol.
3. Transfer buffer: 3.03 g/L Tris base, 14.4 g/L glycine, and 200 mL/L methanol.
4. 3 mm Chromatography paper cut to 9 cm × 7 cm (Whatman).
5. Nitrocellulose membrane (Millipore).
6. PBS: 137 mM NaCl, 2.7 mM KCl, 10 mM Na_2HPO_4, and 2 mM KH_2PO_4.
7. Skimmed milk powder (Anlene™).
8. Antibodies against epitopes or protein of interest.
9. Chemiluminescence substrate kit (Millipore).
10. X-film.

2.6. In Vivo and Semi-In Vivo Protein Degradation Assay

1. HA antibody (Santa Cruz).
2. TRNzol (Tiangen, China).
3. Primers for RT-PCR.
4. ATP (Sigma).

2.7. In Vitro Ubiquitination Assay

1. Ubiquitin (AtUBQ14, At4g02890) with 6× His tag is expressed in *Escherichia coli* using the pET28a vector, wheat E1 (GI: 136632) is expressed using the pET3a vector, and human E2 (UBCh5b) is expressed using the pET15b vector with 6× His tag.
2. The constructs of ubiquitin, E1, and E2 are transformed in *E. coli* strain BL21 (DE3).
3. Proteins are expressed at 18°C for 12–16 h inducing by IPTG and cells were harvested by centrifugation.
4. Pellets are resuspended in lysis buffer and sonicated on ice five times, for 10 s each time.
5. The lysate is centrifuged and supernatant is collected.
6. Ubiquitin, E1, and E2 are quantified by comparing to various amount of BSA protein on SDS-polyacrylamide gel electrophoresis (PAGE).
7. Crude extracts are used for in vitro ubiquitination assay.

3. Methods

Here, we describe an efficient method to study ubiquitination in vivo by transiently expressing putative E3 and substrate protein in *N. benthamiana*. First, the genes encoding E3 or putative substrate to be studied are cloned into plant T-DNA expression vectors as fusions to the 35S constitutive promoter and epitope tags. Each vector is transformed into *A. tumefaciens* strain EHA105 separately, and the *A. tumefaciens* strains carrying the constructs for expressing the substrate and E3 are infiltrated separately or coinfiltrated into leaves of *N. benthamiana* to transiently express the proteins. The interaction between the E3 and substrate is analyzed by coimmunoprecipitation; the ubiquitination of substrate is detected by immunoprecipitation followed by immunoblotting with antibodies against the epitope tag and the ubiquitin; and the effect of E3 ligase on substrate degradation is evaluated by immunoblot or by semi-in vivo protein degradation assays. Proteins expressed by this protocol can also be used for in vitro ubiquitination analysis.

3.1. Growth of Nicotiana benthamiana

1. Mix soil and vermiculite together at a ratio of 1:1, put the mixture into 9 × 9-cm pots, and then water it.
2. Sow about 30 *N. benthamiana* seeds into a pot. Grow plants in a chamber at 22°C and 70% relative humidity under a 16-h light/8-h dark photoperiod.
3. After about 10 days, seedlings are transferred to new 9 × 9-cm pots, one seedling per pot.
4. Plants grown for another 1–1.5 months are used for infiltration (see Note 1).

3.2. Constructs and Bacteria Strains

1. The genes encoding E3, substrate, HAGFP, and *p19* are cloned into plant expression vectors downstream of the 35S constitutive promoter and fused to epitope tags (different tags for E3 and substrate) for detection by antibodies (see Note 2).
2. The overexpression constructs and the empty vector for the E3 construct are transformed into *A. tumefaciens* strains EHA105 separately.

3.3. Agroinfiltration

1. A single colony of *Agrobacterium* is inoculated into 5 mL of LB medium supplemented with the appropriate antibiotics and grown at 28°C in a shaker for 48 h.
2. The culture is transferred to new LB medium with antibiotics, 10 mM MES (pH 5.6), and 40 μM AS (1:100 ratio, *v*/*v*). Agrobacteria are grown at 28°C overnight.
3. When growth reaches an OD_{600} of approximately 3.0, the Agrobacteria are collected by centrifugation at 4,000 × *g*, 10 min, and the pellets are resuspended in 10 mM $MgCl_2$ at a final OD_{600} of 1.5 ($OD_{600} = 1$ for p19 only) (see Note 3).
4. A final concentration of 200 μM AS is added and the Agrobacteria are kept at room temperature for at least 2 h without shaking.
5. Equal volume suspensions of Agrobacteria carrying the construct expected to be expressed and Agrobacteria carrying *p19* are mixed together (see Note 4).
6. Leaf infiltration is conducted by pressing a 1-mL disposable syringe to the surface of fully expanded *N. benthamina* leaves, and then slowly pressing the plunger. Infiltrated leaves have a water-soaked appearance (see Note 5). Put the plants back into the growth chamber for 3 days (see Note 6).

3.4. Sample Collection and Protein Extraction

1. Harvest infiltrated leaves and freeze the tissues in liquid nitrogen as soon as possible.
2. Samples are ground in liquid nitrogen and resuspended in extraction buffer (1 mL/0.2 g sample) by vortex. Use native buffer for immunopreciptation and denaturing buffer for proteins directly analyzed by Western blot (see Note 7).

3. The extracts are centrifuged at 16,000 × *g* at 4°C for 30 min. The supernatant is then collected for immunopreciptation or protein gel blot analysis.

3.5. Immunoprecipitation

1. Extract proteins from tissues using the native buffer supplemented with 50 μM MG132.
2. Add antibody to the cell lysates (10 μg antibody per milliliter cell lysates).
3. Keep the mixture at 4°C with gentle shaking for 2 h.
4. Add 20 μL/mL protein G agarose beads to the mixture and shake at 4°C for another 2 h to capture the immunocomplex (see Note 8).
5. Collect the agarose beads by centrifugation at 14,000 × *g* for 5 s and wash beads with cold PBS three times.

3.6. Western Blotting

1. Add 1/3 volume of 4× SDS sample buffer to the samples and boil for 5 min.
2. Separate proteins by SDS-PAGE in a 10% acrylamide gel and electroblot the gel to nitrocellulose membrane with transfer buffer at 100 V for 75 min (see Note 9).
3. Block the membrane with PBS containing 5% of skimmed milk powder for 1 h at room temperature or overnight at 4°C.
4. Incubate the membrane with primary antibody diluted in PBS containing 3% skimmed milk powder for 1 h at room temperature, and then wash the membrane with 30 mL of PBS twice, 15 min each time.
5. Incubate the membrane with secondary antibody diluted in PBS containing 3% skimmed milk powder for another 1 h at room temperature, and wash with 30 mL PBS three times, 15 min each time.
6. Develop the blots using the Millipore chemiluminescence HRP substrate kit (see Note 10).

3.7. E3 Ligase and Substrate Interaction

1. The *Agrobacterium* strains carrying E3 ligase, substrate, and *p19* genes are used in this assay.
2. The E3 ligase and substrate protein are expressed in *N. benthamiana* separately to avoid degradation of the substrate, following the agroinfiltration procedures described in Subheading 3.3.
3. Samples are collected and extracted with the native buffer.
4. Then, the extract of substrate, the extract of the E3 ligase, and the mixture of both extracts are used for immunoprecipitation (see Note 11).

5. The antibody against the tag of the substrate or of the E3 ligase is used for immunoprecipitation.
6. The antibodies for the tags linked to the substrate and to the E3 ligase are used for Western blot separately to detect their coimmunoprecipitation (interaction).

3.8. Detect Polyubiquitination of Substrate

1. The *Agrobacterium* strains carrying substrate, E3 ligase, and *p19* expression vector are used in this assay.
2. Express the substrate alone or together with E3 ligase by agroinfiltration of *N. benthamiana*. Make extracts with native buffer.
3. Immunoprecipitate the substrate protein using the antibody against its epitope tag.
4. Perform duplicate immunoblots and probe with the antibodies against the tag of the substrate protein and of the ubiquitin to detect the ubiquitinated form of the substrate protein.

3.9. In Vivo Protein Degradation

1. The *Agrobacterium* strains carrying the E3 ligase, substrate, empty vector, HA-GFP, and *p19* in combinations are used in this assay.
2. The *Agrobacterium* strains grow to appropriate OD and are resuspended in 10 mM $MgCl_2$ solution for coinfiltration.
3. Equal volume suspensions of the *Agrobacterium* strains are mixed to express the substrate alone (cotransform with empty vector) or together with the E3 ligase prior to infiltration (see Note 12). *Agrobacterium* strains carrying HA-GFP is also coinfiltrated as an internal control.
4. Three days after infiltration, collect samples for Western blot analysis.
5. Substrate protein is detected by antibody against its epitope tag; the HA-GFP protein level detected by HA antibody is used as an internal control. Expression of the E3 ligase should be confirmed by antibody corresponding to the tag linked to the E3 ligase.
6. The level of substrate proteins accumulated in the (E3 ligase + substrate) sample and in the (vector + substrate) sample are compared to determine whether the degradation of substrate is accelerated by the E3 ligase.
7. The mRNAs of the (E3 ligase + substrate) and the (vector + substrate) samples are extracted by TRNzol. The expressions of both substrate gene and *HAGFP* in both samples are detected by RT-PCR to confirm that the effect of E3 ligase on substrate is indeed at protein level.

3.10. Semi-In Vivo Protein Degradation

1. The *Agrobacterium* strains carrying E3 ligase, substrate, and the empty vector and *p19* genes are used in this assay.
2. Express the substrate protein and E3 ligase separately in *N. benthamiana*. Make extracts using the native buffer.
3. Mix the extract of substrate with that of E3 ligase or vector control in a volume ratio of 1:1 (see Note 11).
4. A final concentration of 10 mM ATP was added to the cell lysates for the function of the 26S proteasome.
5. Incubate the mixtures at 4°C or room temperature (25°C) with gentle shaking (see Note 13).
6. Remove samples at different time points, for example 1, 2, 4, 6, 8, and 12 h for 4°C treatment and 10, 30, 60, 90, 120, and 180 min for 25°C treatment. Stop the reaction by the addition of SDS sample buffer and boil for 5 min (see Note 8).
7. Substrate protein is detected by Western blot using the antibody against the epitope tag. The substrate degradation kinetics in the (E3 ligase + substrate) sample and the (vector + substrate) sample are compared to detect if the E3 ligase promotes the degradation of substrate in the mixture.
8. E3 ligase protein also needs to be detected by antibody corresponding to the tag linked to the E3 ligase.

3.11. The Effect of MG132 on Protein Stability

To detect if the substrate is degraded by the 26S proteasome, we used the 26S proteasome inhibitor, MG132.

1. For in vivo protein degradation, 50 μM MG132 dissolved in 10 mM $MgCl_2$ is infiltrated into the previous infiltrated region 12 h before sample collection.
2. For semi-in vivo degradation, 50 μM MG132 is added into the sample mixture at the beginning of incubation.

3.12. In Vitro Ubiquitination Assay

1. The *Agrobacterium* strains carrying E3 ligase, substrate, and *p19* are used in this assay.
2. Express the substrate and E3 ligase in *N. benthamiana* by agroinfiltration. Mix the extracts of substrate and E3 ligase together and then the mixture is subjected for immunoprecipitation using the antibody against the tag linked to the substrate (see Note 11).
3. About 20 ng of E1, 40 ng of E2, and 5 μg of ubiquitin are added to the tube with immunoprecipitated complex of E3 ligase and substrate; H_2O is added to a final volume of 30 μL. Two control reactions without either E1 or E2 are also mixed.
4. Reactions are carried out at 30°C with agitation in an Eppendorf Thermomixer for 1.5 h. The reactions are stopped by adding SDS sample buffer and boiling for 5 min.

5. Nickel–nitrilotriacetic acid agarose conjugated to horseradish peroxidase is applied for the detection of His-tagged ubiquitin, and antibody against the tag linked to the substrate is used to detect the substrate and its modified form.

4. Notes

1. The growth status of *N. benthamiana* is important. Flowering or unhealthy plants have low transfection efficiency and give a poor protein expression.
2. It has been found that different binary vectors give different protein expression level; when protein is hard to be detected, try different vectors.
3. The Agrobacteria growth to about $OD_{600} = 3$ will give a better protein expression in agroinfilrtation.
4. Gene-silencing suppressor *p19* has been shown to enhance protein expression level greatly. So coinfiltrating protein construct with *p19* expression cassette is strongly recommended, but it is not necessary for those easily expressed proteins.
5. The upper leaves of the *N. benthamiana* plant are better than the lower leaves for protein expression. Plants grow up to eight- to ten-leaves stage and third–fifth leaves from top are the best for agroinfiltration.
6. Generally, we advise to collect sample 3 days after infiltration. However, expression levels of protein varied widely at different postinoculation days for different proteins. Some may have the best expression level at the first day after infiltration and others may peak at the fifth day. A pretest of the time course to collect sample is necessary when the protein expression level is too low to be detected at 3-days point.
7. The native buffer is not suitable for membrane proteins' isolation. To extract membrane protein, a little amount of NP-40 should be added (no more than 1% according to the proteins' property).
8. Large numbers of proteins are very unstable. So if leaving them at 4°C for long time, they might be totally degraded and no signal will be detected. Thus, before immunoprecipitation or protein interaction assay, it is better to take a time course assay at 4°C to determine the protein stability and the best time points used in semi-in vivo protein degradation assay.
9. For proteins of big size, the polyubiquitination form of them may be too big to get into the separating gel and the signals only could be detected in the stacking gel. So when the protein is bigger than 100 kDa, it is better to keep the stacking gel and transfer the proteins in the stacking gel to the membrane.

10. If the background of the Western blot film is high, add 0.1% Tween 20 to PBS at the membrane washing step.
11. The E3 ligase and the substrate protein frequently have different protein expression levels. Thus, for semi-in vivo protein degradation, E3 ligase and substrate interaction, and in vitro ubiquitination assay, the sample volume of E3 ligase and substrate should be adjusted according their protein expression levels.
12. Similar with Note 11, when the E3 ligase expression is too strong, the original bacteria volume of the E3 ligase should be reduced and if the expression level of substrate is much more than the E3 ligase, the original bacteria volume of the substrate should be reduced.
13. For quickly degraded or unstable proteins, we suggest that the incubation should be carried out at 4°C.

Acknowledgments

This research was supported by Grant CNSF31030047/90717006 from the National Natural Science Foundation of China and 973 Program 2011CB915402, which is a grant from National Basic Research Program of China. QX is supported by grants from the Chinese Academy of Science.

References

1. Hershko A, Ciechanover A (1998) The ubiquitin system. Annu Rev Biochem 67:425–479
2. Downes B, Vierstra RD (2005) Post-translational regulation in plants employing a diverse set of polypeptide tags. Biochem Soc Trans 33:393–399
3. Dreher K, Callis J (2007) Ubiquitin, hormones and biotic stress in plants. Ann Bot 99:787–822
4. Hellmann H, Estelle M (2002) Plant development: regulation by protein degradation. Science 297:793–797
5. Fang S, Jensen JP, Ludwig RL, Vousden KH, Weissman AM (2000) Mdm2 is a RING finger-dependent ubiquitin protein ligase for itself and p53. J Biol Chem 275:8945–8951
6. Liu X, Yue Y, Li B, Nie Y, Li W, Wu WH, Ma L (2007) A G protein-coupled receptor is a plasma membrane receptor for the plant hormone abscisic acid. Science 315:1712–1716
7. Chen G, Huang H, Frohlich O, Yang Y, Klein JD, Price SR, Sands JM (2008) MDM2 E3 ubiquitin ligase mediates UT-A1 urea transporter ubiquitination and degradation. Am J Physiol Renal Physiol 295:F1528–F1534
8. Lee HK, Cho SK, Son O, Xu Z, Hwang I, Kim WT (2009) Drought stress-induced Rma1H1, a RING membrane-anchor E3 ubiquitin ligase homolog, regulates aquaporin levels via ubiquitination in transgenic Arabidopsis plants. Plant Cell 21:622–641
9. Liu L, Zhang Y, Tang S, Zhao Q, Zhang Z, Zhang H, Dong L, Guo H, Xie Q (2010) An efficient system to detect protein ubiquitination by agroinfiltration in *Nicotiana benthamiana*. Plant J 61:893–903
10. Voinnet O, Rivas S, Mestre P, Baulcombe D (2003) An enhanced transient expression system in plants based on suppression of gene silencing by the p19 protein of tomato bushy stunt virus. Plant J 33:949–956
11. Zheng N, Xia R, Yang C, Yin B, Li Y, Duan C, Liang L, Guo H, Xie Q (2009) Boosted expression of the SARS-CoV nucleocapsid protein in tobacco and its immunogenicity in mice. Vaccine 27:5001–5007

Chapter 13

In Vitro Protein Ubiquitination Assay

Qingzhen Zhao, Lijing Liu, and Qi Xie

Abstract

Ubiquitination is one of the most important posttranslational modifications in all eukaryote organisms. Ubiquitin-activating enzyme (E1), ubiquitin-conjugating enzyme (E2), and ubiquitin ligase (E3) are the three key enzymes in this process. To detect the specificity between E2 and E3 or enzyme–substrate relationship between E3 and a substrate protein, ubiquitination activity needs to be determined. This protocol provides a convenient and efficient in vitro assay for DTT-sensitive thioester formation of E2s and Ring/U-box-type E3s, and E3-mediated substrate ubiquitination. E2/E3 specificities can also be investigated quickly by using this system. This method can be applied to ubiquitination assays of proteins from any eukaryotic organisms.

Key words: Ubiquitination, Ubiquitin-conjugating enzyme, Ubiquitin ligase, In vitro assay, E2/E3 specificity

1. Introduction

Ubiquitin proteasome system (UPS) is one of the most important regulatory mechanisms in eukaryote cells. UPS seems to play more important roles in plants than in other organisms, as there are a large number of UPS components in plant genomes (1, 2). Attaching the 76 amino acid protein tag ubiquitin to a substrate protein needs the concerted actions of the ubiquitin-activating enzyme (E1), ubiquitin-conjugating enzyme (E2), and ubiquitin ligase (E3). There are 2 E1s, 37 E2s, 8 ubiquitin-conjugating enzyme variants (UEVs), and more than 1,300 E3s in Arabidopsis (3, 4). E1 is considered to have stable and common activities (5). All 37 E2 proteins have a conserved catalytic domain named UBCc domain, which contains a highly conserved Cysteine residue as the binding site of ubiquitin during the ubiquitination process. The eight UEV proteins also have the UBCc domain but lack the conserved Cysteine. The E3 proteins can be divided into

Zhi-Yong Wang and Zhenbiao Yang (eds.), *Plant Signalling Networks: Methods and Protocols*,
Methods in Molecular Biology, vol. 876, DOI 10.1007/978-1-61779-809-2_13,

single-subunit E3s and multisubunit E3s (1, 2, 6, 7). Since the single-subunit E3 protein can carry out its E3 function without the help of other proteins, the ubiquitination activity of this type of E3 can be detected much easily than the multisubunit E3s. The single-subunit E3s contain a conserved Ring finger, U-box, or HECT domain (8). To analyze the function of a putative E2 or E3 protein, its in vitro ubiquitination activity needs to be detected. Up to now, the in vitro system reported in literature often includes components originated from different organisms, such as yeast or rabbit E1, human E2 (H5b or H5c, commonly used), and ubiquitin from different sources (8–11). Due to heterogenous components used, these systems sometimes produce false-positive or -negative results. In this chapter, we provide an in vitro ubiquitination system for E2 and E3 autoubiquitination and E3/substrate ubiquitination assay. All the components of this system, including E1, E2, E3, and Ub, are originally from plants, especially from Arabidopsis. So this system is more compatible for the ubiquitination assay for proteins from plants.

Because a large number of E2/E3-specific interactions have been reported (7, 12–15), different E2/E3 combinations used in the ubiquitination assay will undoubtedly influence the final results. The 37 Arabidopsis E2s can be sorted into 12 groups according to their structure and sequence similarities (1). At least 1 member of each of the 12 E2 groups is tested against an E3 activity to ensure the reliable positive or negative result of the E3 protein. At the same time, E2/E3 specificity can be easily explored. E2 or E3 activities of proteins from other species and ubiquitin modification of a substrate protein by a certain E3 can also be detected.

2. Materials

2.1. Bacteria and Plasmids for Protein Expression

1. *Escherichia coli* strain BL21 (DE3).
2. Plasmid for wheat E1 and human E2 (as control): pET3a-wheat E1 (GI: 136632); pET15b-UBCH5b.
3. Plasmid for Arabidopsis Ub: *pET28a-UBQ14* (At4g02890).
4. Plasmid for Arabidopsis E1: *pET28a-UBA2* (At5g06460).
5. Plasmids for Arabidopsis E2s:
   Coding sequences (CDSs) of at least 1 member of the 12 subfamilies of Arabidopsis E2 (UBC) genes are cloned into the pET28a vector, including *UBC27* (At5g50870), *UBC1* (At1g14400), *UBC2* (At2g02760), *UBC3* (At5g62540), *UBC10* (At5g53300), *UBC32* (At3g17000, delete TM domain), *UBC13* (At3g46460), *UBC4* (At5g41340), *UBC5* (At1g63800), *UBC6* (At2g46030), *UBC21* (At5g25760), *UBC19* (At3g20060), *UBC35* (At1g78870), *UBC16*

(At1g75440), *UBC26* (At1g53020), *UBC22* (At5g05080), and UBC domain of *UBC24* (At2g33770).

2.2. Protein Expression and Purification in E. coli

1. LB medium: Trytone 10 g/L, yeast extract 5 g/L, and NaCl 10 g/L. Autoclaved and store at room temperature.
2. Isopropylthio-β-galactoside (IPTG): 100 mM stock. Dissolve in H_2O (see Note 1), and filter with 0.45-μM sterile membrane. Store at −20°C.
3. Dithiothreitol (DTT): 1 M stock. Dissolve in 10 mM NaAc, and filter with 0.45-μM sterile membrane. Store at −20°C.
4. PMSF: 100 mM stock. Dissolve in isopropanol and store at −20°C.
5. Lysis buffer A (for 6× His-tagged protein expression): 50 mM NaH_2PO_4, 300 mM NaCl, and 1 mM PMSF. Adjust the pH to 8.0. Store at 4°C (see Note 2).
6. Binding buffer (for 6× His-tagged protein purification): 50 mM NaH_2PO_4, 300 mM NaCl, 20 mM imidazole, and 1 mM PMSF. Adjust the pH to 8.0. Store at 4°C.
7. Washing buffer (for 6× His-tagged protein purification): 50 mM NaH_2PO_4, 300 mM NaCl, 50 mM imidazole, and 1 mM PMSF. Adjust the pH to 8.0. Store at 4°C.
8. Elution buffer 1 (for 6× His-tagged protein purification): 50 mM NaH_2PO_4, 300 mM NaCl, 100 mM imidazole, and 1 mM PMSF. Adjust the pH to 8.0. Store at 4°C.
9. Elution buffer 2 (for 6× His-tagged protein purification): 50 mM NaH_2PO_4, 300 mM NaCl, 200 mM imidazole, and 1 mM PMSF. Adjust the pH to 8.0. Store at 4°C.
10. Elution buffer 3 (for 6× His-tagged protein purification): 50 mM NaH_2PO_4, 300 mM NaCl, 300 mM imidazole, and 1 mM PMSF. Adjust the pH to 8.0. Store at 4°C (see Note 3).
11. Lysis buffer B (Column buffer, for MBP-tagged protein expression): 200 mM NaCl, 20 mM Tris–HCl, pH 7.4, 1 mM EDTA, 1 mM DTT, and 1 mM PMSF. Store at 4°C (see Note 2).
12. Beads for 6× His fusion protein purification: Ni-NTA agarose (QIAGEN).
13. Beads for MBP fusion protein purification: Amylose resin (NEB).
14. Centrifugal size-fraction filter: Amicon Ultra-15 (Millipore).

2.3. In Vitro Ubiquitination Assay

1. ATP: 1 M stock. From Sigma. Dissolve in H_2O. Store at −20°C (see Note 4).
2. Reaction buffer (20×) for E2 DTT-sensitive thioester bond formation: 1 M Tris–HCl, pH 7.4, 200 mM $MgCl_2$, and 200 mM ATP. Store at −20°C (see Note 4).

3. Reaction buffer (20×) for E2/E3 and E3/substrate ubiquitination assay: 1 M Tris–HCl, pH 7.4, 200 mM $MgCl_2$, 100 mM ATP, and 40 mM DTT. Store at −20°C (see Note 4).
4. Crude extract of wheat E1 and human E2 expressed in *E. coli* and purified Arabidopsis Ub, E1, E2, and E3 proteins. Store at −80°C.
5. SDS protein sample buffer with β mercaptoethanol (4×): 0.25 M Tris–HCl, pH 6.8, 8% SDS, 40% glycerol, 0.004% bromophenol blue, and 20% β mercaptoethanol. Store at 4°C.
6. SDS protein sample buffer without β mercaptoethanol (4×): 0.25 M Tris–HCl, pH 6.8, 8% SDS, 40% glycerol, and 0.004% bromophenol blue. Store at 4°C.
7. A shaker with constant temperature: For example, thermomixer comfort (Eppendorf).

2.4. SDS-Polyacrylamide Gel Electrophoresis

1. Thirty percent acrylamide/bis-acrylamide (29:1) solution and *N,N,N,N′*-tetramethyl-ethylenediamine (TEMED). Store at 4°C.
2. Ammonium persulfate (APS): Dissolve 1 g of APS in 10 ml H_2O (10% in w/v) and aliquot it in small volume. Store at −20°C.
3. Separating gel buffer (4×): 1.5 M Tris–HCl (pH 8.8) and 0.4% SDS. Store at room temperature.
4. Stacking gel buffer (4×): 1.0 M Tris–HCl (pH 6.8) and 0.4% SDS. Store at room temperature.
5. Running buffer (10×): 0.25 M Tris–HCl, 1.92 M glycine, and 1% SDS. Store at room temperature.
6. Prestained molecular weight marker.

2.5. Western Blotting

1. Transfer buffer: 25 mM Tris–HCl, 192 mM glycine, and 20% (v/v) methanol. Store at room temperature.
2. PBS and PBST buffer: Prepare 10× PBS stock with 1.37 M NaCl, 27 mM KCl, 100 mM Na_2HPO_4, and 20 mM KH_2PO_4. Adjust the pH to 7.4. Dilute 100 ml of stock with 900 ml H_2O for the working buffer (1× PBS). For 1× PBST buffer, add 0.5 ml Tween-20 into 1 L 1× PBS solution. Store at room temperature.
3. Blocking buffer: 5% nonfat dry milk in 1× PBST.
4. Primary and secondary antibody dilution buffer: 3% nonfat dry milk in 1× PBS.
5. Primary antibody: Anti-His antibody (Santa Cruz).
6. Secondary antibody: Anti-mouse IgG conjugated with horseradish peroxidase (HRP) (Santa Cruz).
7. HRP substrate kit (Millipore).

3. Methods

Expression in bacteria system is the most convenient way for us to obtain the proteins of candidate genes. Although some proteins of eukaryote genes expressed from *E. coli* are inactive due to the lack of posttranslational modification or other reasons, most of them do have activities and can be applied for further analysis (8, 16). Recombined proteins with a certain tag can be easily purified according to the corresponding methods. In order to avoid the disturbance of the components in elution buffer of purification, the purified proteins need to be desalted by Amicon ultra centrifugation to remove the high concentration of salt or sugars and so on. Then, the proteins can be applied for in vitro ubiquitination assay. In this chapter, E2 (DTT-sensitive thioester assay), E2/E3 specificity, and E3/substrate ubiquitination assay are described.

3.1. Expression, Purification of Proteins in Bacteria

1. Protein expression in bacteria: Plasmids are transformed in *E. coli* strain BL21 (DE3) (see Note 5). Select two or three clones in 5 ml of LB liquid medium and shake at 37°C. The overnight culture is then diluted with LB containing 0.2% glucose to generate a start OD_{600} absorbance of 0.1 and subsequently shaken at 18°C (see Note 6). Cells are induced with a final concentration of 0.2 mM IPTG when the OD_{600} reaches 0.4–0.6. After induction, the cells are kept on growing for 12–16 h at 18°C. Cells are harvested by centrifugation. The collected cells for 6 × His-tagged proteins are resuspended in lysis buffer A and MBP fusion protein cells in lysis buffer B (column buffer). The cells are then disrupted by ultrasonic pulverizor. The cell lysate is centrifuged at 13,500 × g at 4°C for 45 min and the supernatant is taken into a new tube.
2. Purification of 6× His-tagged protein: The Ni-NTA agarose is packed into column (see Note 7). Wash the beads with distilled water. Equilibrate the column with ten-column volumes of the lysis buffer. Then, let the supernatant sample containing the recombined proteins flow through the column (see Note 8). Wash the column with five-column volumes of washing buffer. Elute the proteins with elution buffer 1, 2, and 3 in turn (2 ml each) (see Notes 3 and 9).
3. Purification of MBP-tagged protein: Pour the amylase resin into column (see Note 7). Load the supernatant of cell lysate into the column (see Note 8). Wash the column with 12-column volumes of column buffer. Elute the protein with column buffer containing 10 mM maltose. Collect ten fractions of 1 ml each in turn. Usually we can get enough purified proteins in the first five fractions.
4. Purification of proteins by Amicon centrifuge columns: Dilute the purified proteins (in elution buffer) with PBS buffer into

total volume of 15 ml. Load the solution to a water-prewashed centrifuged filter (Amicon ultra 15). Centrifuge at 4°C, 4,000 × *g*, until the remaining volume is about 0.2–0.5 ml (see Note 10). The protein can be checked by quantification and SDS-PAGE electrophoresis. Glycerol is added to a final concentration of 15–20%. Then, aliquot the protein, and store at −80°C (see Note 11).

3.2. E2 Autoubiquitination Assay

1. The reaction is performed in total 30 μl, including 1.5 μl of 20× buffer, 50 ng of E1, 200–500 ng E2, and 2 μg ubiquitin.
2. Incubate the reactions at 37°C for 5 min.
3. Split the reactions by adding 4× SDS sample buffer with or without β-mercaptoethanol (see Note 12).
4. Boil the samples at 100°C for 5 min.
5. The reaction products are separated with 12% SDS-PAGE gel and detected with anti-His antibody by Western blotting to detect the DTT-sensitive thioester bonds.

3.3. Combined E2/E3 Specificity Assay

The E3 proteins can be purified in a 1.5-ml Eppendorf tube from cell crude extract just before use (see Notes 13 and 14).

1. Wash the amylase resin beads with 1 ml of column buffer.
2. Centrifuge at 400 × *g* for 2 min. Remove the liquid carefully.
3. Add 0.5–1 ml of crude extract (the total amount of the E3 protein should be 0.5–1 μg) to the tube containing the pre-washed beads.
4. Rotate at room temperature for 1 h for binding.
5. Wash the beads with 1 ml of 50 mM Tris–HCl (pH 7.5) for three times and remove all the liquid of the final wash using very thin tips (see Note 15).
6. Prepare the reactions in total 30 μl, including 1.5 μl of 20× reaction buffer, 50 ng of E1, 200–500 ng of E2, and 5 μg of ubiquitin. Add the reaction system to the tubes containing the amylase resin beads binding with MBP-E3 proteins (see Note 16).
7. Incubate the reactions at 30°C for 1.5 h with agitation in a thermomixer (Eppendorf) (see Note 16).
8. Split the reactions by adding 4× SDS sample buffer with β-mercaptoethanol and boil the samples at 100°C for 5 min.
9. The reaction products are separated with 10–12% SDS-PAGE gel and detected with anti-His antibody by Western blotting.

3.4. E3/Substrate Ubiquitination Assay

1. Clone the substrate genes into the TNT vector.
2. Express the substrate protein using the TNT–substrate construct and wheat germ in vitro translation system (see Note 17).

3. Detect the expression effect by 1–3 μl of translation products (see Note 18).
4. Bind the E3 proteins to the beads (see steps 1–5 of Subheading 3.3 and Note 13).
5. Add the tube with the following reagents: 1.5 μl of 20× reaction buffer, 50 ng of El, 200–500 ng of E2, 3–8 μl of TNT substrate mix, and 5 μg of ubiquitin.
6. Incubate the reactions at 30°C for 1.5 h with agitation in a thermomixer (Eppendorf) (see Note 16).
7. Add 10 μl of 4× protein-loading buffer with β-mercaptoethanol and boil the samples at 100°C for 5 min.
8. The reaction products are separated with 10–12% SDS-PAGE gel and detected with anti-His antibody by Western blotting.

3.5. Western Blotting

1. Proteins resolved by SDS-PAGE gel are transferred to nitrocellulose membrane with transfer buffer at 100 V for 75 min.
2. Block the membrane with blocking buffer for at least 1 h at room temperature or overnight at 4°C.
3. Remove the blocking buffer; incubate the membrane with 1:300-fold diluted primary antibody (anti-His) in antibody dilution buffer for 1 h at room temperature.
4. Remove the primary antibody and wash the membrane with PBST buffer twice, 15 min each.
5. Incubate the membrane with secondary antibody (goat-anti-mouse IgG conjugated with HRP) diluted in antibody dilution buffer (1:2,500-fold diluted) for 1 h at room temperature.
6. Wash the membrane with PBST buffer twice, 15 min each.
7. Detect the signals with the Millipore chemiluminescent HRP substrate kit.
8. For the E3/substrate assay with substrate from the in vitro translational system, the ubiquitination of the substrate can be directly detected by exposing the gel to the X-ray film.
9. For the E3/substrate assay with purified substrate from *E. coli*, the ubiquitination of the substrate can be detected by Western blotting with the antibody corresponding to the tag fused with the substrate.

4. Notes

1. The "H_2O" in this text refers to the ultrapurified water having a resistivity of 18.2 MΩ-cm and total organic content of less than five parts per billion.

2. Stocks of DTT (1 M) and PMSF (100 mM) are not stable at 4°C or room temperature; thus, they are stored at −20°C. They should be added to the solution as working concentration just before use. PMSF is very unstable in aqueous solutions. The 100 mM stock solution is made in isopropanol.
3. The only difference among three types of elution buffers is the concentration of imidazole in solution. We recommend that three different elution buffers be tested for the first-time trial to check in which elution buffer you will recover most of the particular proteins you are analyzing.
4. ATP is inactivated rapidly at room temperature. So the stock of ATP and ubiquitination reaction buffer containing ATP should be aliquoted into small volume and stored at −20°C. It should be used as soon as dissolved and should not be frozen and thawed repeatedly.
5. *E. coli* BL21 (DE3) is a commonly used bacterial strain for protein expression. In case some proteins cannot be induced well in this strain because of the rare code, toxicity protein, and so on, you may try other strains, such as BL21 (AI) or BL21 (plys S) and so on.
6. We recommend growing the diluted culture at 18°C because the low temperature is helpful for the solubility of the target protein which is important for further purification process. Higher temperature, such as 37 or 25°C, is also fit if the protein can be easily obtained from the supernatant after centrifugation of the cell lysate. This can be tried by the advanced experiment. Glucose is advantageous to the induction of the protein driven by T7 promoter.
7. The amount of resin depends on the fusion protein produced and the binding capacity of the beads (usually, 1 ml beads can bind 2–3 mg proteins).
8. The flow rate should be about 1 ml/min.
9. High purity of the imidazole is needed for this process. The concentrations of the imidazole in binding buffer, washing buffer, and elution buffer are most important for the purification efficiency. To obtain the highest purity of a certain protein, the optimal concentration should be first determined by using a gradient concentration of imidazole.
10. There are different types of the Amicon centricon. The appropriate type number for a certain protein should be less than 1/3 of the molecular weight of that protein.
11. All the manipulation of the protein should be on ice and as soon as possible in order to reserve its maximal activity. Usually, proteins are more stable at −80°C than −20 or 4°C. Protein activity will decline when left in 4°C for a period of time. So the

proteins should be aliquoted into small volumes, and stored at −80°C until usage. Freezing and thawing repeatedly should be avoided.

12. At the presence of El and ubiquitin, active E2 proteins can be linked with ubiquitin through thioester bonds. These thioester bonds can be broken off at the presence of the reductive agent DTT or β-mercaptoethanol. So the E2 autoubiquitination activity can be proved by this DTT-sensitive thioester bonds.
13. Purified E3 proteins treated by Amicon centricon can also be used for this assay.
14. It has been proved that there exists a large amount of E2/E3 specificity. So different E2/E3 combinations can result in different activities. Use this system E3 activity of one certain protein can be explored by combined with different E2s (for example, it may exhibit E3 activity at the present of one E2 but not together with another E2). So the E2/E3 specificity can also be explored with this system. MBP proteins and reactions without El or E2 are necessary as negative control for the E3 activity assay.
15. The operation with the amylase resin beads must be done carefully to avoid taking out the beads unexpectedly. After discarding the supernatant following washing the beads, the tube can be spinned shiefly, and the left liquid can be moved with very thin tips. With this action we can eliminate the wash buffer completely and it is helpful for the following operation.
16. In order to ensure the effect of the ubiquitination reaction, the beads must be covered by the total volume of the reaction and resuspended enough by shaking in the mixer.
17. Purified recombined substrate proteins from *E. coli* fused with different tags with the E3 can also be used in this test. But in some cases, this kind of substrate cannot be ubiquitinated by the E3 because of lack of posttranslational modification; for example, phosphorylation is important for a large amount of substrate proteins.
18. The translation products must be sufficient enough because the ubiquitination form of the substrate always occupies a little percent of the total proteins.

Acknowledgments

This research was supported by Grant CNSF31030047/90717006 from the National Natural Science Foundation of China and 973 Program 2011CB915402, which is a grant from National Basic Research Program of China. Q.X. is supported by grants from the Chinese Academy of Science.

References

1. Bachmair A, Novatchkova M, Potuschak T, Eisenhaber F (2001) Ubiquitylation in plants: a post-genomic look at a post-translational modification. Trends Plant Sci 6:463–470
2. Vierstra RD (2003) The ubiquitin/26S proteasome pathway, the complex last chapter in the life of many plant proteins. Trends Plant Sci 8:135–142
3. Smalle J, Vierstra RD (2004) The ubiquitin 26S proteasome proteolytic pathway. Annu Rev Plant Biol 55:555–590
4. Vierstra RD (2009) The ubiquitin-26S proteasome system at the nexus of plant biology. Nat Rev Mol Cell Biol 10:385–397
5. Hatfield PM, Gosink MM, Carpenter TB, Vierstra RD (1997) The ubiquitin-activating enzyme (E1) gene family in *Arabidopsis thaliana*. Plant J 11:213–226
6. Kerscher O, Felberbaum R, Hochstrasser M (2006) Modification of proteins by ubiquitin and ubiquitin-like proteins. Annu Rev Cell Dev Biol 22:159–180
7. Ye Y, Rape M (2009) Building ubiquitin chains: E2 enzymes at work. Nat Rev Mol Cell Biol 10:755–764
8. Stone SL, Hauksdottir H, Troy A, Herschleb J, Kraft E, Callis J (2005) Functional analysis of the RING-type ubiquitin ligase family of Arabidopsis. Plant Physiol 137:13–30
9. Bu Q, Li H, Zhao Q, Jiang H, Zhai Q, Zhang J, Wu X, Sun J, Xie Q, Wang D, Li C (2009) The Arabidopsis RING finger E3 ligase RHA2a is a novel positive regulator of abscisic acid signaling during seed germination and early seedling development. Plant Physiol 150:463–481
10. Xie Q, Guo HS, Dallman G, Fang S, Weissman AM, Chua NH (2002) SINAT5 promotes ubiquitin-related degradation of NAC1 to attenuate auxin signals. Nature 419:167–170
11. Zhang Y, Yang C, Li Y, Zheng N, Chen H, Zhao Q, Gao T, Guo H, Xie Q (2007) SDIR1 is a RING finger E3 ligase that positively regulates stress-responsive abscisic acid signaling in Arabidopsis. Plant Cell 19:1912–1929
12. Christensen DE, Klevit RE (2009) Dynamic interactions of proteins in complex networks: identifying the complete set of interacting E2s for functional investigation of E3-dependent protein ubiquitination. FEBS J 276:5381–5389
13. Kraft E, Stone SL, Ma L, Su N, Gao Y, Lau OS, Deng XW, Callis J (2005) Genome analysis and functional characterization of the E2 and RING-type E3 ligase ubiquitination enzymes of Arabidopsis. Plant Physiol 139:1597–1611
14. Markson G, Kiel C, Hyde R, Brown S, Charalabous P, Bremm A, Semple J, Woodsmith J, Duley S, Salehi-Ashtiani K, Vidal M, Komander D, Serrano L, Lehner P, Sanderson CM (2009) Analysis of the human E2 ubiquitin conjugating enzyme protein interaction network. Genome Res 19:1905–1911
15. van Wijk SJ, de Vries SJ, Kemmeren P, Huang A, Boelens R, Bonvin AM, Timmers HT (2009) A comprehensive framework of E2-RING E3 interactions of the human ubiquitin-proteasome system. Mol Syst Biol 5:295
16. Wiborg J, O'Shea C, Skriver K (2008) Biochemical function of typical and variant *Arabidopsis thaliana* U-box E3 ubiquitin-protein ligases. Biochem J 413:447–457

Chapter 14

Genome-Wide Identification of Transcription Factor-Binding Sites in Plants Using Chromatin Immunoprecipitation Followed by Microarray (ChIP-chip) or Sequencing (ChIP-seq)

Jia-Ying Zhu, Yu Sun, and Zhi-Yong Wang

Abstract

Nearly all signal transduction pathways lead to regulation of gene expression by controlling specific transcription factors (TFs). Chromatin immunoprecipitation (ChIP) is a powerful method for studying TF–DNA interactions in vivo. To identify all binding sites of a TF in the genome, the DNA obtained in ChIP experiments needs to be analyzed by hybridization to genome-tiling microarrays (ChIP-chip) or by next-generation sequencing (ChIP-seq). Here, we provide detailed protocols of ChIP for two model plant species Arabidopsis and rice, procedures of DNA sample preparation for ChIP-chip or ChIP-seq, and a general guide for computational data analysis. We have used these protocols to successfully identify direct target genes of the BZR1 TF of the brassinosteroid signaling pathway in both Arabidopsis and rice.

Key words: Chromatin immunoprecipitation, ChIP-chip, ChIP-seq, Transcription factors

1. Introduction

All signal transduction pathways in plants regulate gene expression through specific transcription factors (TFs), which bind to cis-elements in or near the target genes. Genome-wide identification of the binding sites of the TFs of signal transduction pathways is an effective approach to uncovering all direct target genes controlled by the signaling pathway. This is important not only for understanding the functions of signaling pathways, but also for elucidating the transcription networks integrating signaling pathways. Chromatin immunoprecipitation (ChIP), first described by Varshavsky and colleagues (1), is an effective method for studying in vivo protein–DNA interactions (1–3). ChIP experiment starts with covalently cross-linking a protein to its bound DNA by formaldehyde (1).

Zhi-Yong Wang and Zhenbiao Yang (eds.), *Plant Signalling Networks: Methods and Protocols*, Methods in Molecular Biology, vol. 876, DOI 10.1007/978-1-61779-809-2_14,

The chromatin is then extracted and DNA is fragmented before the protein–DNA complex is immunoprecipitated by specific antibody against the DNA-binding protein. The purified DNA can be analyzed by quantitative real-time PCR (ChIP-qPCR) to determine the enrichment of genomic regions of interest. Alternatively, the DNA can be analyzed by hybridization to genome-tiling microarray (ChIP-chip) or by high-throughput sequencing (ChIP-seq) to identify all binding sites in the genome (Fig. 1). ChIP-chip and ChIP-seq are the most powerful methods for identifying the genomic targets of TFs (4).

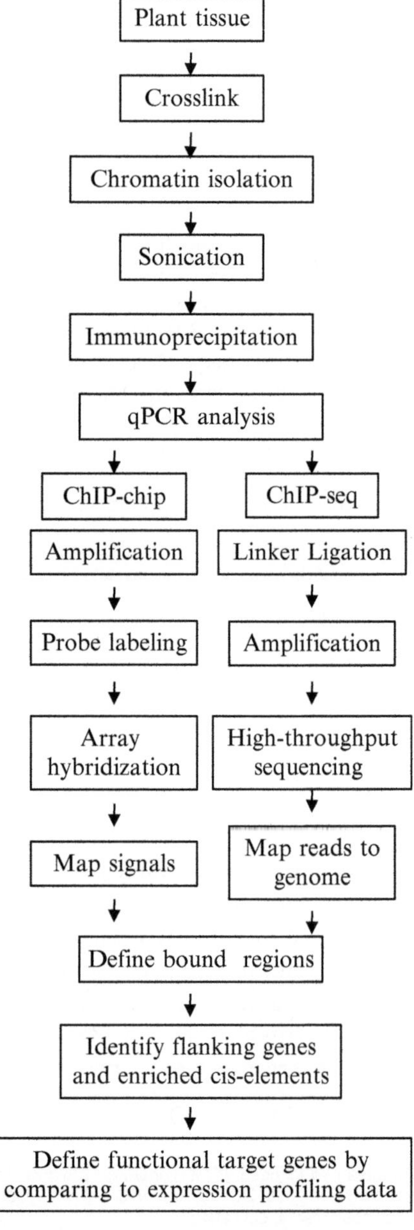

Fig. 1. Workflow of ChIP-chip and ChIP-seq. This outline presents the procedure of ChIP-chip and ChIP-seq experiments and the following data analysis.

ChIP-seq has some advantages over ChIP-chip in several aspects. ChIP-chip requires whole genome microarray, which is available for only a limited number of model species, whereas ChIP-seq produces the sequences of the bound DNA that only need to be aligned to the reference genome. In addition, microarray has limited sensitivity, specificity, and dynamic range; thus, ChIP-seq has greater resolution, sensitivity, and specificity than ChIP-chip (5, 6). With the cost of sequencing dropping and technology being improved, ChIP-seq has become the method of choice. However, ChIP-chip is still a valid method when access to sequencing facility is limited (5, 7). Both ChIP-chip and ChIP-seq have been widely used in the research of humans and animal systems (8), but only a few studies have been reported for plants (9–16). We have successfully identified several thousands of direct target genes of BZR1, the key TF of the brassinosteroid signaling pathway in Arabidopsis using ChIP-chip (17), and targets of OsBZR1 in rice using ChIP-seq (Zhu et al., unpublished results).

For ChIP-seq, several massively parallel sequencing platforms are available currently, including Illumina Genome Analyzer (GA), 454 Life Science, Helicos HeliScope, and the Applied Biosystem SOLID System (18). Illumina Genome Analyzer is most widely used for ChIP-seq. In this protocol, we provide methods for performing ChIP-seq using Illumina platform. In general, the Illumina platform can provide more than 10^8 reads per sample lane. The new Illumina's HiSeq 2000 Genome Analyzer increases output to 600 Gb per run and up to six billion paired-end reads (http://www.illumina.com/). We advise using the latest version available and bar-coded library and multiplex sequencing runs to reduce cost and time (19). While the number of targets identified increases with the depth of sequencing, the number of Pol II targets identified in human cells saturates at about ten million reads (20).

Moving from raw ChIP-seq data to biological insight requires sequential data analysis (Fig. 1). Many open sources and tools based on different methods are available for ChIP-chip and ChIP-seq data analysis, and have been discussed in detail in recent reviews (19). In this protocol, we describe important factors to consider when designing ChIP-chip and ChIP-seq experiments, and we provide detailed protocol for ChIP and DNA sample preparation for ChIP-chip and ChIP-seq analyses in Arabidopsis and rice, but only brief guidelines for data analysis.

2. Materials

2.1. Plant Tissue Cross-Link

1. Cross-link buffer: 1% formaldehyde.
2. Solution for stopping cross-link: 0.25 M glycine.

2.2. Chromatin Extraction Buffer

1. Extraction buffer 1 (EB1): 0.4 M sucrose, 10 mM Tris–HCl (pH 8.0), 5 mM β-mercaptoethanol, 1 mM PMSF, and Protease Inhibitor cocktail (see Note 1).
2. Extraction buffer 2 (EB2): 0.25 M sucrose, 10 mM Tris–HCl (pH 8.0), 5 mM β-mercaptoethanol, 10 mM $MgCl_2$, 1% Triton X-100, 1 mM PMSF, and Protease Inhibitor cocktail.
3. Extraction buffer 3 (EB3): 1.7 M sucrose, 10 mM Tris–HCl (pH 8.0), 5 mM β-mercaptoethanol, 2 mM $MgCl_2$, 0.15% Triton X-100, 1 mM PMSF, and Protease Inhibitor cocktail.
4. 2× Nuclei lysis buffer (NLB): 10 mM Tris–HCl (pH 8.0), 20 mM EDTA, 400 mM NaCl, 1% Triton X-100, 2 mM PMSF, and Protease inhibitor cocktail.

2.3. Immunoprecipitation

1. Protein A-magnetic beads or sheared salmon sperm DNA/protein A agarose beads (Upstate Biotechnology) (see Note 2).
2. Antibody against the protein of interest.
3. High-salt wash buffer: 50 mM Tris–HCl (pH 8.0), 2 mM EDTA, 500 mM NaCl, and 0.25% Triton X-100.
4. LiCl wash buffer: 10 mM Tris–HCl (pH 8.0), 1 mM EDTA, 25 mM LiCl, 0.5% NP-40, and 0.25% sodium deoxycholate.
5. Elution buffer: 1% SDS and 0.1 M $NaHCO_3$. Prepare fresh before use.
6. TE buffer: 10 mM Tris–HCl (pH 8.0) and 1 mM EDTA.

2.4. ChIP-DNA Recovery

1. Reverse cross-linking of DNA and protein complex: 5 M NaCl.
2. Remove protein and RNA: 14 mg/ml proteinase K, RNase A, and 1 M Tris–HCl (pH 6.5).
3. DNA extract: Phenol/chloroform/isoamyl alcohol (25:24:1), 8 M LiCl, 100% and 70% ethanol, and 1 mg/ml glycogen.

2.5. ChIP-seq DNA Library

1. Illumina adaptors and PCR primers (Illumina).
2. End-It DNA end repair kit (Epicentre).
3. 10 mM dNTP stock (NEB).
4. Klenow enzyme (NEB).
5. T4 DNA ligase with 10× buffer (Promega).
6. Gel extraction kit (QIAGEN).
7. Zero blunt TOPO PCR cloning kit for sequencing (Invitrogen).
8. Sanger sequencing reagents.

2.6. ChIP-chip DNA Preparation

1. Whole Genome Amplification Kit (WGA1) (Sigma).
2. Taq DNA Polymerase (Fermentas).
3. QIAquick PCR Purification Kit (QIAGEN).

4. DNaseI (Invitrogen).
5. Affymetrix GeneChip Arabidopsis Tiling 1.0R array.
6. GeneChip WT Double-Stranded DNA Terminal Labeling Kit (Affymetrix).
7. GeneChip® Hybridization, Wash, and Stain Kit (Affymetrix).

3. Methods

3.1. Experimental Design

To ensure high-quality data in genome-wide analysis, the choices of antibody and negative control must be carefully considered. First, the antibody must specifically immunoprecipitate the protein of interest. Cross-reaction with other chromatin-associated proteins generates false-positive data, although this problem can be alleviated by including proper negative controls. Polyclonal antibodies are easier to make and are usually more effective in immunoprecipitation than monoclonal antibodies, while monoclonal antibodies usually have higher specificity. ChIP can be performed using the antibody against the TF itself in wild-type plants or using antibody to the epitope tag, such as GFP, MYC, or FLAG tags, fused to the TF expressed in transgenic plants.

An advantage of using antibodies against the TF itself is that natural situation is analyzed in wild-type plants. Epitope-tagged protein may not behave the same as the native protein due to interference by the tag or altered expression pattern of the transgene. However, it is difficult to design a perfect negative control when using anti-TF antibody. Often, preimmune serum or no antibody is used as negative control; but false positives can occur if the immune serum cross-reacts with unrelated proteins or homologs of the TF. We have made antibody using maltose-binding protein (MBP)-tagged recombinant OsBZR1, and then purified anti-OsBZR1 IgG using GST-OsBZR1 and purified control IgG using MBP. However, even antigen affinity-purified antibodies can cross-react with nonspecific proteins, as we often detect in Western blot. Knockout mutant of the TF can also be used as negative control, but false negatives can occur if any homologous proteins bind to the same DNA when the specific TF is absent and are recognized by the antibody. Some mutants have strong effects on development and change tissue composition of the samples, which may indirectly affect the results.

When transgenic plants expressing epitope-tagged protein are used for ChIP, nontransgenic plants or plants expressing the tag only (large tags, such as GFP) provide ideal negative controls. Ideally, the tagged TF should be expressed under the TF's own promoter in the knockout mutant background to confirm that the tagged TF has normal function and can complement the mutant and to avoid competition for DNA binding by the untagged endogenous TF.

Overexpression of the TF using a strong constitutive promoter can potentially increase sensitivity but can also lead to false positives.

For ChIP-chip or ChIP-seq, the negative-control ChIPs often yield insufficient amount of DNA (the cleaner the ChIP experiment, the less DNA recovered in the negative control), and additional or scaled-up ChIP experiments need to be performed to accumulate enough control DNA for sequencing or array hybridization. Input DNA is often analyzed as negative control for ChIP-seq and ChIP-chip. ChIP-seq has been also performed without negative control, and some software identify binding sites based on relative enrichment of different chromosomal regions. However, it has been observed that some chromosomal regions, such as regions of open chromatin structure, can be enriched nonrandomly in the input DNA (20). Therefore, negative control (control IgG, nontransgenics for tagged proteins, or input DNA) seems essential for ensuring high-confidence ChIP-seq data.

The number of biological replicates is another factor to be considered when designing ChIP-chip and ChIP-seq experiments. Biological replicates ensure that experiments are reproducible and the variation between samples in an experiment can be quantified. According to Rozowsky et al. (20), three biological replicates are required for ChIP-chip experiments and two replicates are sufficient for ChIP-seq experiments if there is over 90% agreement between the replicates. The third replicate only marginally improves the data quality of ChIP-seq experiments. Pooling DNA from multiple ChIP experiments before probe synthesis or library construction is an economic way to improve data quality.

3.2. Antibody–Beads' Preparation

For each ChIP assay using 3 g of tissue, wash 30 μl of ssDNA/protein A-agarose beads or protein A-magnetic beads (see Note 2) with 0.3× NLB buffer three times. Spin at 1,000 × *g* for 1 min and keep one volume of supernatant with beads. Add 5 μg of antibodies, mix, and incubate at 4°C on a rotator overnight or longer until immunoprecipitation experiment.

3.3. Tissue Cross-Linking

1. Harvest plant tissues of interest, and rinse with water to wash off the dirt. If using rice seedlings as plant materials, the tissues need to be cut into small pieces (see Note 3).
2. Merge tissue into 1% formaldehyde. Vacuum for 5 min, turn off the vacuum, then release the gas, and hold for 5 min (10 min for rice). Vacuum infiltration is not necessary for root tissues or etiolated seedlings.
3. To quench cross-linking reaction, transfer tissues to a 0.25 M glycine solution.
4. Rinse seedlings three times with deionized water. Remove surface water from tissues using paper towels.
5. Freeze tissues in liquid nitrogen and keep in −80°C freezer or use immediately (see Note 4).

3.4. Chromatin Isolation

All the centrifuge and spin steps are performed at 4°C. Samples are always kept on ice or 4°C cold room and all buffers used are precooled unless otherwise indicated.

1. Grind the cross-linked tissue in liquid nitrogen into fine powder. Weigh 3 g of tissue powder into a 50-ml tube frozen in liquid nitrogen.
2. Add 30 ml of cold buffer EB1 to 3 g of tissue powder in 50-ml tube, vortex, and mix thoroughly.
3. Filter solution through two layers of Miracloth into a new ice-cold 50-ml tube.
4. Centrifuge solution for 10 min at 1,800 × g.
5. Remove supernatant and resuspend pellet in 1.5 ml of EB2 buffer, and transfer the solution to a 1.5-ml microcentrifuge tube.
6. Centrifuge at 20,000 × g for 10 min.
7. Remove supernatant. If the pellet is still green, repeat steps 5 and 6 up to two times or until the pellet is light green (see Note 5).
8. Fully resuspend the pellet in 500 μl of EB3 buffer. Layer the solution onto another 500 μl of EB3 in a new 1.5-ml microcentrifuge tube and centrifuge at 20,000 × g for 1 h at 4°C.
9. Remove supernatant and resuspend pellet (chromatin) in 250 μl of 2× NLB buffer. Save 5 μl for DNA gel analysis (see Note 6).
10. Sonicate the chromatin solution for 6 × 15 pulses (for ChIP-seq) or 4 × 15 pulses (for ChIP-chip) (for rice, double the sonication time), 1 min break between each 15 pulses, 40% duty, and output level 4 using a micro probe (Branson Sonifier) to achieve an average DNA size of 200 bps (for ChIP-seq) or 0.5–1.5 kb (for ChIP-chip). Care should be taken to avoid foaming. Keep the tube cooled in ice water all the time.
11. Centrifuge the chromatin sample for 10 min, 20,000 × g, at 4°C. Transfer the supernatant to a new tube and centrifuge for 5 more minutes. Take aliquot of 5 μl of supernatant to check sonication efficiency. The remaining supernatant can be kept at −80°C freezer or used immediately for immunoprecipitation.
12. To check sonication efficiency (see Note 7), reverse cross-link by adding equal volume of 0.5 M NaCl and heat the chromatin extract at 95°C for 30 min. Run a 2% agarose gel to determine the average size of DNA fragments compared with the aliquot from step 9 above.

3.5. Immunoprecipitation

1. Add 1 ml of 0.3× NLB buffer to antibody–protein A beads that are prepared in Subheading 3.2. Split equally in two tubes for control and sample if control tissue is used (no antigen). Spin and remove liquid.

2. Add 2× volume of ddH_2O to the sheared chromatin solution. Save 10 μl to use as input for Western blot in step 8 below. Save 50 μl if input is to be used as control for ChIP-qPCR or ChIP-seq.
3. Add the supernatant to antibody/Protein A beads (of step 1), and incubate on a rotator from 1 h to overnight at 4°C.
4. Spin at 1,000 × *g* for 1 min. Save supernatant (flow through) for Western blot.
5. Wash beads with 1 ml of 0.5× NLB buffer, invert tube five times, and spin at 1,000 × *g* for 30 s to remove supernatant. Repeat wash (see Note 8).
6. Wash beads with 1 ml of high-salt wash buffer (one wash), 1 ml of LiCl buffer (one wash), and 1 ml of TE buffer (two washes; transfer beads to a new 0.5-ml tube during the first TE wash). These washing steps are performed as follows: fully resuspend the beads by adding the buffer, then spin immediately at 1,000 × *g* for 1 min, and use a 1-ml pipette to remove the supernatant.
7. Elute the immunoprecipitated protein–DNA complex with 250 μl of prewarmed (65°C) elution buffer and incubate at 65°C for 15 min. Spin to collect elute and repeat elution of beads.
8. Combine the two elutes, add 20 μl of 5 M NaCl, and reverse the cross-linking at 65°C overnight. Save 10 μl to check the efficiency of immunoprecipitation by immunoblotting.
9. Add 10 μl of 0.5 M EDTA, 20 μl of 1 M Tris–HCl (pH 6.5), 1.5 μl of 14 mg/ml proteinase K, and 1 μg of RNase A to the combined eluate and incubate at 45°C for 1 h (see Note 9).
10. Purify DNA using a commercial PCR purification kit. Alternatively, extract DNA by equal volume of phenol/chloroform/isoamyl alcohol (25:24:1) and precipitate DNA with LiCl/ethanol containing glycogen at −20°C overnight (for 500 μl of supernatant, add 5 μl of 1 mg/ml glycogen, 50 μl of 8 M LiCl, and 1.1 ml of 100% ethanol). Wash pellets with 70% ethanol.
11. Resuspend the precipitated DNA in 15 μl of TE buffer for ChIP-chip purpose or 50 μl in terms of ChIP-seq. Measure DNA concentration using Picogreen assay (Invitrogen Q-bit) and Agilent BioAnalyzer DNA 1000 chip. Store at −20°C for the next analysis.

3.6. qPCR Analysis

Before moving on to microarray hybridization or sequencing, the ChIP DNA should be analyzed by qPCR to confirm enrichment of known target genes relative to nontarget control genes. Alternatively, if no known target is available for positive control, the amounts of immunoprecipitated DNA can be compared to that of the negative control immunoprecipitation that uses no antibody or a sample without the antigen. Relative enrichment of target genes can be calculated as ratio (ΔCT of qPCR) of sample to the negative control or to input DNA, and the data should be normalized to a nontarget control gene, such as the 25S rDNA. We used at least two nonbinding regions, such as *CNX5* and *UBC30*, for BZR1, one for data normalization, and the other(s) as negative control (17). Primer pairs should amplify about 200 bps of target or nontarget control regions of the genome. DNAs from TF ChIP and negative controls [either no-antibody (NoAb) or no-antigen controls, or input DNA] are analyzed for each primer pair used.

1. Dilute part of the DNA sample 1:5 in milliQ water (per reaction: 1 μl of sample to 5 μl final volume).
2. Prepare a master mix of forward and reverse primers at a final concentration of 1 μM per primer.
3. Combine the following in a PCR tube:

Primer master mix	0.5 μl
milliQ water	3.5 μl
Diluted DNA sample	1 μl
SYBR Green mix	5 μl

4. Set up and run the following PCR program: initial denaturation for 2 min at 95°C, followed by a total of 40 cycles of denaturation at 95°C for 15 s, anneal at 60°C for 15 s, extend at 72°C for 20 s, and then for 5 min at 72°C.
5. ChIP-qPCR data analysis:
   Fold enrichment for a gene is calculated as the ratio of TF ChIP to control ChIP (NoAb or no-antigen control). Relative fold enrichment for target or test genes is normalized to a nontarget control gene.
   (a) Calculate fold enrichment in the sample: $2^{-(\text{CT IP} - \text{CT negative control})}$.
   (b) Calculate relative fold enrichment:
   Fold enrichment of target gene/fold enrichment of control gene.

Upon qPCR confirmation of high fold enrichment (greater than fivefold) of known target genes, the ChIP-enriched DNA can be used for either ChIP-seq or ChIP-chip.

3.7. ChIP-seq

3.7.1. Construction of DNA Library for Illumina Sequencing

Use at least 10 ng of ChIP-enriched DNA for an Illumina ChIP-seq library. Pooling DNA from multiple immunoprecipitation reactions is recommended to achieve this quantity and to minimize incidental variation. Some sequencing facilities or services take the ChIP-enriched DNA and prepare library for sequencing; but if you plan to perform multiple ChIP-seq experiments, it is usually less expensive if you make the library yourself. Using bar-code and multiplexing sequencing runs can further reduce cost. This is recommended particularly when using the newer HiSeg2000 analyzer, which can generate hundreds of millions of reads per run. The following protocol is for preparation of library for sequencing by Illumina HiSeq.

1. End repair: Combine and mix the following components in a tube, and incubate at room temperature (18–20°C) for 45 min:

ChIP-DNA	10 ng
10× and repair buffer	5 μl
2.5 mM dNTP	5 μl
10 mM ATP	5 μl
End-repair enzyme mix	1 μl

   Water to bring reaction volume to 50 μl.
2. Purify DNA using PCR purification columns and elute DNA in 34 μl of elution buffer.
3. A-tailing: Combine and mix the following components in a PCR tube and incubate for 30 min at 37°C:

DNA from step 2	34 μl
Klenow buffer	5 μl
1 mM dATP	10 μl
Klenow fragment (3′–5′ exonuclease minus)	1 μl
Total reaction	50 μl

4. Purify DNA using PCR purification columns and elute DNA in 17 μl of elution buffer.
5. Adaptor ligation: Combine and mix the following components in a tube and incubate for 20–22 h at 16°C (see Note 10):

DNA from step 4	16.5 μl
10× DNA ligase buffer	2 μl
Illumina adaptor oligo mix (1:10–1:50)	1 μl
T4 DNA ligase	0.5 μl
Total reaction volume	20 μl

6. Purify DNA using PCR purification kit and elute DNA in 30 μl of elution buffer.

7. Combine and mix the following components in a PCR tube:

5× Phusion buffer	10 μl
2.5 mM dNTP	4 μl
Phusion enzyme	0.8 μl
ddH_2O	3.2 μl
DNA from step 6	30 μl
PCR primer 1.1	1 μl
PCR primer 2.1	1 μl
Total reaction volume	50 μl

8. Amplify using the following PCR procedure: initial denaturation at 98°C for 30 s, followed by a total of 18 cycles of denaturation at 98°C for 10 s, anneal at 65°C for 30 s, and extend at 72°C for 30 s and then for 5 min at 72°C.
9. Purify PCR product and elute DNA in 20 μl of elution buffer.
10. Test for enrichment using qPCR with the same conditions as described in Subheading 3.6.
11. Gel purification: Run PCR product on a 2% agarose gel, leaving empty lanes between samples to avoid cross-contamination. Excise a gel band in the range of 200–500 bps. Purify the DNA from the agarose slice using gel extraction kit. Elute DNA in 30 μl of elution buffer.
12. Test the quality of DNA:
    (a) Test for enrichment using qPCR as described in Subheading 3.6.
    (b) Measure the DNA concentration and size by picogreen assay (Invitrogen Q-bit) and Agilent BioAnalyzer DNA 1000 chip.
    (c) Optional: Use 2 μl of DNA from step 11 for standard ligation into a Topo vector. Transform into *Escherichia coli* (e.g., TOP10) and sequence 10–20 clones by Sanger sequencing. More than 30% of the cloned inserts should match to Arabidopsis or rice genome.
13. Dilute library to 10 nM. Prepare flow cell using 4 pM final concentration by following the Illumina protocol (see Note 11).

3.7.2. Sequencing

Sequence the DNA library using Illumina Genome Analyzer, such as HiSeq2000, and process the data according to instructions of Illumina for ChIP-Seq data analysis.

3.7.3. ChIP-Seq Data Analysis

1. Align the sequence reads to the reference genome. Select the sequences that match uniquely to one place in the genome.

Several programs are currently available for reads mapping, including ELAND (Illumina, http://bioit.dbi.udel.edu/howto/eland), MAQ (http://map.sourceforge.net/index.shtml) (21), and SOAP (http://soap.genomics.org.cn/) (22).

2. Analyze the matched sequences to detect genomic regions with significant enrichment of mapped reads using one of the software for ChIP-seq analysis, such as CisGenome (23), MACS (24), PRI-CAT (25), or PeakSeq (20).

3.8. ChIP-chip

3.8.1. Preparation of DNA Probe for ChIP-chip

If ChIP-seq is not possible and genome-tiling array is available, ChIP-chip is a viable alternative. Compared to ChIP-seq, larger amount of DNA is required to make labeled probes for microarray hybridization in ChIP-chip. ChIP DNA must be amplified by PCR. We recommend using the Whole Genome Amplification kit (Sigma) to amplify the DNA, and the amplified DNA should be analyzed by qPCR to confirm that enrichment of known target genes is maintained after the amplification. We suggest pooling amplified DNA of at least three independent ChIP experiments to accumulate enough DNA for probe synthesis and to minimize variation, as we did for the BZR1 ChIP-chip experiment (17).

1. Take out 10 μl of immunoprecipitated DNA (see Note 12) to a PCR tube, add 1 μl of 10× Fragmentation Buffer, and put immediately back on ice.
2. Add 2 μl of 1× Library Preparation Buffer and 1 μl of Library Stabilization Solution, mix by gently tapping the bottom of the tube, and place the tube in thermal cycler at 95°C for 2 min. Cool sample on ice and spin briefly.
3. Add 1 μl of Library Preparation Enzyme to the same PCR tube, mix thoroughly, and spin briefly.
4. Place the sample in thermal cycler and incubate as follows: 16°C for 20 min, 24°C for 20 min, 37°C for 20 min, and 75°C for 5 min. Take out the tube immediately and put back on ice.
5. Add the following reagents to the entire 15 μl of reaction to make a total volume of 75 μl: 7.5 μl of 10 × Amplification Master Mix, 3 μl of Taq DNA polymerase, and 49.5 μl of nuclease-free water. Mix thoroughly, centrifuge briefly, and put into the preheated thermocycler immediately for PCR amplification.
6. Use the PCR condition as follows: initial denaturation at 95°C for 3 min, then 15 cycles (see Note 13) of denaturation at 94°C for 15 s, anneal/extend at 65°C for 5 min, and hold at 4°C.
7. Purify the PCR product with PCR purification kit. The total yield of amplified DNA should be about 5–10 μg.

8. Perform qPCR to confirm that enrichment of known target gene is similar to preamplification level.
9. Repeat the entire ChIP experiment two more times as biological repeats.

3.8.2. Probe Labeling

1. Take 3.5 μg of amplified DNA from each of the three repeat experiments and combine them together (see Note 14).
2. Fragment the pooled DNA to an average of 50–100 bps with DNase I at 30°C for 10 min (see Note 15).
3. Load 2 μg of the fragmented DNA on a 2% agarose gel to check that the DNA size is about 50–100 bps.
4. Biotin label 7.5 μg of fragmented DNA with TdT following the Affymetrix Chromatin Immunoprecipitation Assay protocol: Mix 7.5 μg of amplified ChIP DNA, 10 μl of 5× TdT buffer, 3.5 μl of $CoCl_2$, 2 μl of biotin-ddATP, and 0.1 μl of Terminal transferase and nuclease-free water to 50 μl. Incubate at 37°C for 1 h. Then, incubate at 95°C for 10 min.

3.8.3. Hybridization, Washing, Staining, and Scanning

Hybridize the labeled DNA to Affymetrix GeneChip Arabidopsis Tiling 1.0R array (see Note 16). Perform microarray hybridization, washing, staining, and scanning steps according to the Eukaryotic Target Protocol (Affymetrix).

3.8.4. ChIP-chip Data Analysis

To identify the statistically enriched regions, microarray hybridization data can be analyzed by one of the available software for ChIP-chip data analysis, such as Tiling Analysis Software (TAS, Affymetrix) and CisGenome (23). The ChIP-chip or ChIP-seq data can be uploaded and visualized with genome browsers, such as the UCSC Genome Browser (http://genome.ucsc.edu/) (26) or Integrated Genome Browser (IGB, Affymetrix; http://www.affymetrix.com/partners_programs/programs/developer/tools/download_igb.affx/) (27).

4. Notes

1. PMSF and other protease inhibitors should be added immediately before the use of buffer.
2. Do not use DNA-coated beads for ChIP-seq experiment because the coating DNA contributes to sequencing reads.
3. This protocol presents general steps of ChIP-seq and ChIP-chip for Arabidopsis. Modifications of the protocol for rice samples are noted in the protocol. For example, we have found that sonication at higher power and/or for longer time is needed to shear rice DNA than Arabidopsis DNA.

4. The cross-link step is critical for ChIP experiment. The efficiency of cross-link can be evaluated using Western blot analysis of cross-linked and reverse-cross-linked samples. Additional bands or smear should be observed close to the top of the gel and the signal of the protein-only band should be reduced compared to cross-link-reversed sample. It is also important not to overcross-link the sample to limit the loss of antibody recognition site and false positive.
5. After EB2 extraction, the pellet should be light green, which means most chloroplasts are lysed. If it is possible, a larger volume of EB2 buffer can be used to avoid repeating steps 5 and 6.
6. At this stage, save a 5 μl of aliquot for later examination. Aliquots taken at this step as well as at step 11 must be treated as described in elution and reverse cross-linking procedure before analyzing on the agarose gel.
7. We recommend checking the size of DNA fragment for each experiment.
8. At this step, the suspended beads can be transferred to a 0.5-ml tube to make it easier to remove wash solution without losing beads.
9. For ChIP-seq, this step is necessary because the high quality of DNA is required for sequencing; while for ChIP-chip, this step is optional.
10. The adaptors should be diluted. The optimal concentration of adaptors needs to be determined empirically.
11. The 10 nM library can be stored at −80°C for several months.
12. Since the precipitated DNA amount is very small, it is not necessary to measure the DNA concentration as noted in the WGA protocol. We typically use as much DNA as possible.
13. The PCR cycle number can be adjusted from 14 to 18 cycles as necessary.
14. Pooling DNA from three amplifications reduces the variation introduced during amplification since the amplification does not amplify the DNA fragments evenly.
15. The DNase I fragmentation condition is variable. A pilot experiment should be done to optimize the condition.
16. For Arabidopsis, Agilent and Nimblegen tiling arrays are also available.

Acknowledgment

This work was supported by National Institutes of Health Grant R01GM066258 (to Z.-Y. W.).

References

1. Solomon MJ, Larsen PL, Varshavsky A (1988) Mapping protein–DNA interactions in vivo with formaldehyde: evidence that histone H4 is retained on a highly transcribed gene. Cell 53:937–947
2. Kuo MH, Allis CD (1999) In vivo cross-linking and immunoprecipitation for studying dynamic protein:DNA associations in a chromatin environment. Methods 19:425–433
3. Orlando V (2000) Mapping chromosomal proteins in vivo by formaldehyde-crosslinked-chromatin immunoprecipitation. Trends Biochem Sci 25:99–104
4. Kim TH, Ren B (2006) Genome-wide analysis of protein–DNA interactions. Annu Rev Genomics Hum Genet 7:81–102
5. Johnson DS, Mortazavi A, Myers RM, Wold B (2007) Genome-wide mapping of in vivo protein–DNA interactions. Science 316:1497–1502
6. Schmidt D, Stark R, Wilson MD, Brown GD, Odom DT (2008) Genome-scale validation of deep-sequencing libraries. PLoS One 3:e3713
7. Barski A, Cuddapah S, Cui K, Roh TY, Schones DE, Wang Z, Wei G, Chepelev I, Zhao K (2007) High-resolution profiling of histone methylations in the human genome. Cell 129:823–837
8. Morozova O, Marra MA (2008) Applications of next-generation sequencing technologies in functional genomics. Genomics 92:255–264
9. Lee J, He K, Stolc V, Lee H, Figueroa P, Gao Y, Tongprasit W, Zhao H, Lee I, Deng XW (2007) Analysis of transcription factor HY5 genomic binding sites revealed its hierarchical role in light regulation of development. Plant Cell 19:731–749
10. Oh E, Kang H, Yamaguchi S, Park J, Lee D, Kamiya Y, Choi G (2009) Genome-wide analysis of genes targeted by PHYTOCHROME INTERACTING FACTOR 3-LIKE5 during seed germination in Arabidopsis. Plant Cell 21:403–419
11. Kaufmann K, Muino JM, Jauregui R, Airoldi CA, Smaczniak C, Krajewski P, Angenent GC (2009) Target genes of the MADS transcription factor SEPALLATA3: integration of developmental and hormonal pathways in the Arabidopsis flower. PLoS Biol 7:e1000090
12. Morohashi K, Grotewold E (2009) A systems approach reveals regulatory circuitry for Arabidopsis trichome initiation by the GL3 and GL1 selectors. PLoS Genet 5:e1000396
13. Kaufmann K, Wellmer F, Muino JM, Ferrier T, Wuest SE, Kumar V, Serrano-Mislata A, Madueno F, Krajewski P, Meyerowitz EM, Angenent GC, Riechmann JL (2010) Orchestration of floral initiation by APETALA1. Science 328:85–89
14. Yant L, Mathieu J, Dinh TT, Ott F, Lanz C, Wollmann H, Chen X, Schmid M (2010) Orchestration of the floral transition and floral development in Arabidopsis by the bifunctional transcription factor APETALA2. Plant Cell 22:2156–2170
15. Thibaud-Nissen F, Wu H, Richmond T, Redman JC, Johnson C, Green R, Arias J, Town CD (2006) Development of Arabidopsis whole-genome microarrays and their application to the discovery of binding sites for the TGA2 transcription factor in salicylic acid-treated plants. Plant J 47:152–162
16. Zhang H, He H, Wang X, Yang X, Li L, Deng XW (2011) Genome-wide mapping of the HY5-mediated gene networks in Arabidopsis that involve both transcriptional and post-transcriptional regulation. Plant J 65:346–358
17. Sun Y, Fan XY, Cao DM, Tang W, He K, Zhu JY, He JX, Bai MY, Zhu S, Oh E, Patil S, Kim TW, Ji H, Wong WH, Rhee SY, Wang ZY (2010) Integration of brassinosteroid signal transduction with the transcription network for plant growth regulation in Arabidopsis. Dev Cell 19:765–777
18. Kaufmann K, Muino JM, Osteras M, Farinelli L, Krajewski P, Angenent GC (2010) Chromatin immunoprecipitation (ChIP) of plant transcription factors followed by sequencing (ChIP-SEQ) or hybridization to whole genome arrays (ChIP-CHIP). Nat Protoc 5:457–472
19. Lefrancois P, Zheng W, Snyder M (2010) ChIP-Seq using high-throughput DNA sequencing for genome-wide identification of transcription factor binding sites. Methods Enzymol 470:77–104
20. Rozowsky J, Euskirchen G, Auerbach RK, Zhang ZD, Gibson T, Bjornson R, Carriero N, Snyder M, Gerstein MB (2009) PeakSeq enables systematic scoring of ChIP-seq experiments relative to controls. Nat Biotechnol 27:66–75
21. Li H, Ruan J, Durbin R (2008) Mapping short DNA sequencing reads and calling variants using mapping quality scores. Genome Res 18:1851–1858

22. Li R, Li Y, Kristiansen K, Wang J (2008) SOAP: short oligonucleotide alignment program. Bioinformatics 24:713–714

23. Ji H, Jiang H, Ma W, Johnson DS, Myers RM, Wong WH (2008) An integrated software system for analyzing ChIP-chip and ChIP-seq data. Nat Biotechnol 26:1293–1300

24. Zhang Y, Liu T, Meyer CA, Eeckhoute J, Johnson DS, Bernstein BE, Nusbaum C, Myers RM, Brown M, Li W, Liu XS (2008) Model-based analysis of ChIP-Seq (MACS). Genome Biol 9:R137

25. Muino JM, Hoogstraat M, van Ham RC, van Dijk AD (2011) PRI-CAT: a web-tool for the analysis, storage and visualization of plant ChIP-seq experiments. Nucleic Acids Res 39: W524–W527

26. Karolchik D, Baertsch R, Diekhans M, Furey TS, Hinrichs A, Lu YT, Roskin KM, Schwartz M, Sugnet CW, Thomas DJ, Weber RJ, Haussler D, Kent WJ (2003) The UCSC Genome Browser Database. Nucleic Acids Res 31:51–54

27. Nicol JW, Helt GA, Blanchard SG Jr, Raja A, Loraine AE (2009) The Integrated Genome Browser: free software for distribution and exploration of genome-scale datasets. Bioinformatics 25:2730–2731

Chapter 15

Smart Pooling of mRNA Samples for Efficient Transcript Profiling

Raghunandan M. Kainkaryam, Angela Bruex, Peter J. Woolf, and John Schiefelbein

Abstract

Gene expression profiling studies are commonly used to study signaling pathways and their impact on transcriptional regulation in plants. In some cases, a profiling study results in expression profiles in which most genes exhibit a small number of differentially expressed states among a large number of samples. In such instances, a pooling approach would help improve the efficiency of the profiling effort by employing fewer microarray chips and ensuring more robust measurement of transcript levels. Smart pooling involves pooling of mRNA samples in an information-efficient manner such that each sample is tested multiple times but always in pools with other samples. The resulting pooled measurements are then decoded to recover the expression profile of all samples in the study. In this protocol, we describe in detail the process of designing smart pooling experiments and decoding their results, which have been used for studying signaling in Arabidopsis root development. Heuristics are provided to select the design parameters that would ensure successful execution of smart pooling.

Key words: Transcriptional regulation, Gene regulatory network, Microarray, Pooling, Sparsity, Compressive sensing, *L1*-Minimization

1. Introduction

Signaling networks often exert their effects by altering the level of transcription of individual genes. Accordingly, the complete understanding of a signaling network ideally requires the definition and analysis of the entire suite of genes that are affected by changes in the activity of the network. Microarray experiments have become a very popular approach to analyze genome-wide changes in transcript accumulation in response to changes in signaling processes. Numerous studies have now been conducted in which changes in transcript profiles have been defined for a variety of plant-signaling processes,

Zhi-Yong Wang and Zhenbiao Yang (eds.), *Plant Signalling Networks: Methods and Protocols*, Methods in Molecular Biology, vol. 876, DOI 10.1007/978-1-61779-809-2_15,

including developmental, hormonal, and environmental (1, 2). More broadly, the Gene Expression Omnibus (GEO) database currently holds more than 500,000 biological samples profiled using genomic profiling technologies from approximately 20,000 studies (3).

In a typical expression profiling study, multiple samples of treated and untreated material are each assayed individually. Given the cost of the gene chips and assay reagents, a design strategy that reduced the number of chips required for a microarray study would be desirable. Furthermore, testing each sample in multiple assays, rather than a single assay, would reduce the impact of noise and increase the robustness of the transcript measurements. We have recently proposed a pooled approach to organizing a transcript profiling experiment, which we term "poolMC," designed to use fewer microarray measurements (i.e., chips) as well as to reduce the measurement noise (4). The central idea of this approach is to employ "smart" pooling of mRNA samples, prior to chip hybridization, such that each biological sample is tested multiple times across the study and always in pools containing other samples. Computational decoding of the resulting pooled measurements would accurately and robustly recover the expression profiles of each sample in the study while using fewer chips than a regular one-sample-one-chip approach. This smart pooling approach is inspired by the new field of compressive sensing that aims to compress data at the measurement source rather than at a later point (5). The following protocol describes the procedure to design a smart pooling experiment and decodes its results. It also provides heuristics to help decide if and when this approach is applicable and useful to gene expression profiling studies using microarray technologies.

2. Materials

2.1. Microarray Experiments

1. Qiagen RNeasy kit (Qiagen, Inc, Germantown, MD) for isolation of mRNA from the plant samples.
2. NuGen Ovation v2 and NuGen Ovation FL kits (NuGen, Inc, San Carlos, CA) for labeling samples.
3. Affymetrix ATH1 Genechip® (Affymetrix, Inc, Santa Clara, CA) for measurement of gene expression profiles of the samples.

2.2. Microarray Data-Preprocessing Software

1. Robust Multiarray Averaging (6) software: http://biosun01.biostat.jhsph.edu/~ririzarr/affy/.
2. Custom CDF from University of Michigan's Brainarray Microarray Lab (7): http://brainarray.mbni.med.umich.edu/Brainarray/Database/CustomCDF/CDF_download.asp.

2.3. Pooling Design and Decoding Software

1. MATLAB® scientific programming software: http://www.mathworks.com/.

2. *L1*-MAGIC linear programming software (8): http://www.acm.caltech.edu/L1magic.
3. Pooling design code (.m file) provided as supplementary material (open access) in ref. 4: http://www.biomedcentral.com/content/supplementary/1471-2105-11-299-s4.m.
4. Pooling result decoder (.m file) provided as supplementary material (open access) in ref. 4: http://www.biomedcentral.com/content/supplementary/1471-2105-11-299-s5.m.

3. Methods

This section describes the procedure to generate the mRNA pooling design, experimentally pool the samples, and computationally decode the results. This approach may be applied to any microarray study, in theory, but it is most robust in instances where gene expression differences are "sparse," that is, where a relatively small number of genes exhibit significant differences in transcript levels between the contrasting experimental conditions. This situation is likely to be common, particularly if specific cell types or tissues are being examined or if a well-defined signaling pathway is being considered.

3.1. Generation of Pooling Design

1. Choose the design parameters (see Note 1) for the pooling experiment: number of samples to be pooled (n), number of chips to use (m), number of times each sample is tested across all m chips (d), and the maximum number of samples that should be pooled on each chip (r).
2. Use the pooling design MATLAB® code (see Subheading 2.3, step 2) with the input parameters n, m, d, and r to generate an $m \times n$ binary matrix containing a 1 for each sample (column) to be pooled in the corresponding chip (row) and a 0 otherwise. An example is shown in Fig. 1. The figure shows a pooling experiment, where 25 samples are pooled as per the design shown into 20 pooled samples that are tested on 20 microarray chips, representing a saving of 20%. The design tests each sample on three different chips, each time pooled with a different set of samples, with at most five samples pooled together on any chip.

3.2. Experimentally Pool Samples

1. Isolate mRNA from plant material (according to the protocol provided by the Qiagen RNeasy kit) and determine its concentration in each sample (see Note 2).

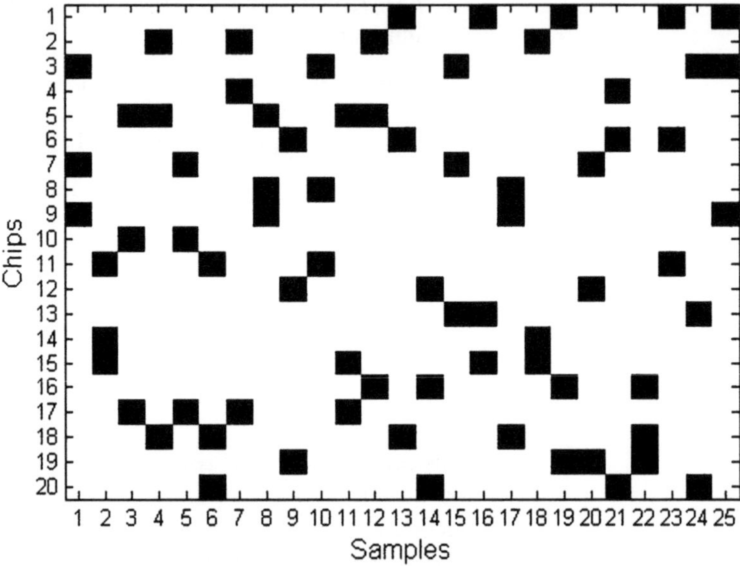

Fig. 1. An example of a pooling design using 25 samples tested on 20 microarray chips. Each *row* represents one mRNA pool that is used for microarray analysis on one chip. Each *column* represents one sample and indicates the contribution of each particular sample into multiple pools. Thus, each *black square* represents the presence of a sample (along that *column*) within the pool tested on a specific chip (along that *row*).

2. It is recommended that a total of approximately 3.5 μg mRNA be used per chip, divided among each of the RNAs contributing to the pool.
3. Mix fractions of RNA samples based on the design generated in Subheading 3.1 (see Note 3).
4. Generate the cDNA and label using the NuGen Ovation v2 and the NuGen Ovation FL kits (following protocols provided in the kits).
5. Hybridize a total of 4 μg of the labeled cDNA on the Affymetrix ATH1 Genechip (following protocols provided at www.Affymetrix.com).

3.3. Preprocessing of Microarray Data

1. Once the microarray chips have been scanned, standard data preprocessing methods, such as Robust Multiarray Averaging (6), can be used to obtain the expression levels of individual probe sets. The resulting data is typically log transformed.
2. The user can choose to use updated probe set definitions (in the form of custom Chip Description Files) provided by repositories, such as the University of Michigan's Brainarray Microarray Lab (7).
3. The resulting $p \times m$ data matrix with probe sets (p) as rows and the pooled experiments (m) as columns is used as input to the poolMC decoder.

3.4. Decode the Expression Levels of Pooled Samples

1. The data matrix described in Subheading 3.3, step 3, and the pooling design described in Subheading 3.1, step 2, are used as inputs to the poolMC decoder (see Subheading 2.3, step 3).
2. The decoder returns the expression profiles for each individual sample in the pooling experiment (see Note 4).

4. Notes

1. The heuristic (described in Fig. 5 of ref. 4) for choosing the design parameters is as follows: for a chosen sparsity of k-spikes in n samples, the number of measurements that guarantee successful recovery scales as $k\log(n/k)$. For example, a study with 100 samples would require approximately 23 chips, if only 1 in 10 samples showed differential expression. However, to account for measurement noise and experimental variability, more chips should be used. Since estimating the sparsity of an experiment a priori is difficult, it is useful to use the following approach. It is reasonable to assume that the number of samples that can be pooled together on a single chip is at most five (due to concentration–detection limits). Further, each sample should be tested in at least two different pools. For example, for a sample size (n) of 100, if each sample is tested in three pools and each pool comprises at most four samples, the number of chips (m) to be used is $100 \times 3/4 = 75$. This represents a 25% savings in the number of chips used over the traditional one-sample-one-chip design, and it assures a more reliable reconstruction of expression profiles of the individual samples.
2. To maximize the robustness of this approach, it is important to reduce the differences between samples as much as possible. For example, it is best to obtain RNA from a specific cell type or tissue, rather than a whole organ or the entire plant. Also, a specific treatment or alteration to a signaling pathway, such as a single gene alteration, is likely to lead to the best results. In our own research, we have made use of fluorescence-based cell sorting with a cell-type-specific GFP marker line [e.g., *WEREWOLF* promoter::green fluorescent protein (GFP) fusion (*WER::GFP*)] to limit the variation in gene expression across the samples (9).
3. The fractional concentration of each sample to be pooled on a given chip is determined by the number of 1s in each row of the pooling design. For a total concentration of 3.5 μg mRNA tested on each chip, each sample in the pool should be mixed at $3.5/r$ μg mRNA, where r is the number of 1s in that row (pool) of the pooling design. This also implies that, as long as the minimum pool size (r) is less than the number of times (d) each

sample is tested across the pooling design, the total (across the whole pooling experiment) mRNA required for each sample would not be more than 3.5 μg.

4. The decoder takes as input the log-transformed $p \times m$ data matrix as input along with the $m \times n$ pooling design matrix. The pooling design is then adjusted to account for the fractional mixtures of samples, simply by dividing each row in the matrix by the number of 1s in that row. On line 59 of the poolMC decoder (see Subheading 2.3, step 3), a parameter *epsilon* is set to 1 and can be increased to reduce sensitivity to noise at the cost of quantitative accuracy. Our experience has been that an *epsilon* of 1 works well for most microarray studies. The decoder solves a linear program for each gene (or probe set) on the chip, namely, each row (p) in the data matrix. The individual decoded expressions for the genes are then accumulated into the resulting decoded expression profile, which is a $p \times n$ matrix. The decoding process for each gene on the chip should require only a few minutes of time (depending on the number of samples in the study) to run on a standard laptop.

Acknowledgments

We thank Christa Barran for technical assistance with this research. This project has been financially supported by an NSF grant (IOS 0744599) to J.W.S.

References

1. Lucas M, Laplaze L, Bennett MJ (2011) Plant systems biology: network matters. Plant Cell Environ. doi:10.1111/j.1365-3040.2010.02273.x
2. Sreenivasulu N, Sunkar R, Wobus U, Strickert M (2010) Array platforms and bioinformatic tools for the analysis of plant transcriptome in response to abiotic stress. Methods Mol Biol 639:71–93
3. Barrett T, Troup DB, Wilhite SE, Ledoux P, Evangelista C, Kim IF et al (2011) NCBI GEO: archive for functional genomics data sets—10 years on. Nucleic Acids Res 39:D1005–D1010
4. Kainkaryam RM, Bruex A, Gilbert AC, Schiefelbein J, Woolf PJ (2010) poolMC: smart pooling of mRNA samples in microarray experiments. BMC Bioinformatics 11:299
5. Candes EJ, Wakin MB (2008) An introduction to compressive sampling. Signal Process Mag IEEE 25(2):21–30
6. Irizarry RA, Hobbs B, Collin F, Beazer-Barclay YD, Antonellis KJ, Scherf U et al (2003) Exploration, normalization, and summaries of high density oligonucleotide array probe level data. Biostat 4(2):249–264
7. Dai M, Wang P, Boyd AD, Kostov G, Athey B, Jones EG et al (2005) Evolving gene/transcript definitions significantly alter the interpretation of GeneChip data. Nucleic Acids Res 33(20): e175–e179
8. Candes EJ, Romberg J (2005) *l1*-MAGIC: recovery of sparse signals via convex programming. http://www.acm.caltech.edu/l1magic
9. Lee MM, Schiefelbein J (1999) WEREWOLF, a MYB-related protein in Arabidopsis, is a position-dependent regulator of epidermal cell patterning. Cell 99:473–483

Chapter 16

Transient Expression Assays for Quantifying Signaling Output

Yajie Niu and Jen Sheen

Abstract

The protoplast transient expression system has become a powerful and popular tool for studying molecular mechanisms underlying various plant signal transduction pathways. *Arabidopsis* mesophyll protoplasts display intact and active physiological responses and are easy to isolate and transfect, which facilitate high-throughput screening and systematic and genome-wide characterization of gene functions. The system is suitable for most *Arabidopsis* accessions and mutant plants. Genetic complementation of mutant defective in sensor functions, gene expression, enzymatic activities, protein interactions, and protein trafficking can be easily designed and explored in cell-based assays. Here, we describe the detailed protocols for protoplast isolation, polyethylene glycol-calcium transfection, and different assays for quantifying the output of various signaling pathways.

Key words: Arabidopsis, Mesophyll protoplast, Transient expression, Signal transduction, Reporter assay, Genetic complementation, Genomics

1. Introduction

As sessile organisms, plants have evolved sophisticated mechanisms to regulate their growth and development as well as to adjust their physiological state for survival in response to environmental changes and challenges. Perception of the environmental cues and subsequent signal transduction lead to the transcriptional and physiological reprogramming that are central to these adaptive responses. Although genetic approach has been extensively used to identify signal transduction components (1–4), the functional redundancy or overlap in the *Arabidopsis* genome and the dynamics and complexity of intertwined signaling mechanisms make it difficult to understand the role of a particular gene, often with multiple unrelated functions. Gain-of-function approaches, such as generating overexpression transgenic plants, can help to elucidate the function of the gene of interest; however, they are more time consuming and

Zhi-Yong Wang and Zhenbiao Yang (eds.), *Plant Signalling Networks: Methods and Protocols*,
Methods in Molecular Biology, vol. 876, DOI 10.1007/978-1-61779-809-2_16,

unpredictable or yield puzzling and complex results due to long-term gene expression consequences that are remote from the original gene activity. Transient gene expression in isolated plant cells provides a versatile and rapid alternative approach, which is a relatively simple way to explore new ideas and investigate the role of a candidate gene.

Since the isolation of plant protoplasts was reported 50 years ago (5), protoplasts from various plant species have been widely employed in physiological, biochemical, and molecular studies, such as signal transduction processes, ion transport, cell wall synthesis, protein trafficking, viral replication, and programmed cell death (1). *Arabidopsis* mesophyll protoplasts isolated from fresh leaves have been proven to be ideal for studying signal transduction pathways (1, 6, 7), as they respond sensitively and specifically to diverse signals, such as heat, cold, darkness, light, elicitors, H_2O_2, auxin, abscisic acid (ABA), and cytokinin (Fig. 1). Moreover, fast and simple procedures

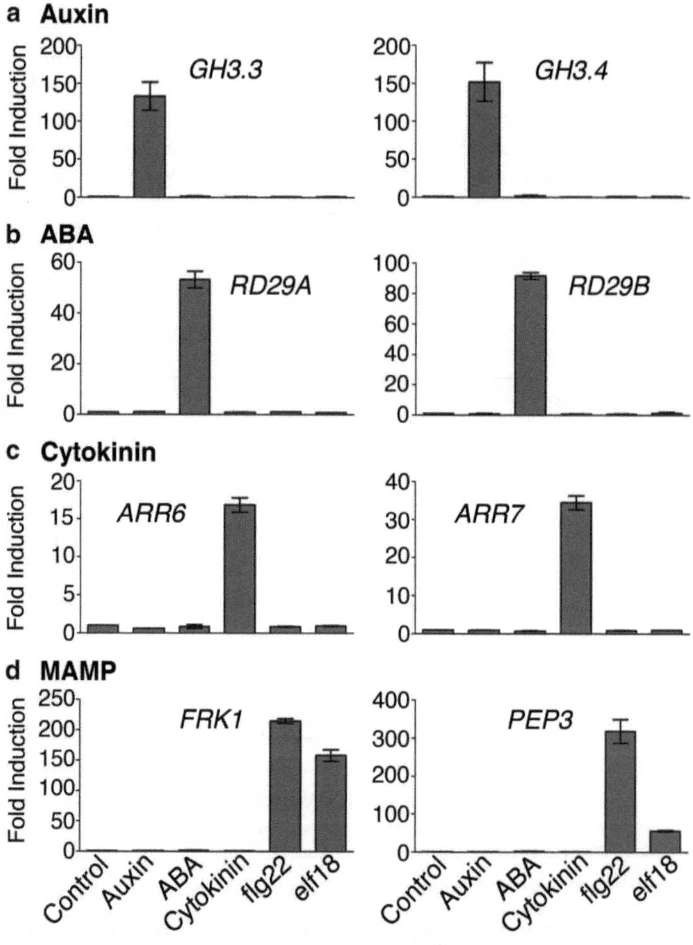

Fig. 1. Quantitative real-time RT-PCR analysis of marker gene expression in response to different signals. The protoplasts were treated for 1 h with 10 μM NAA (auxin), 10 μM ABA, 100 nM 2-iP (cytokinin), 10 nM flg22, or 100 nM elf18. Protoplasts without treatment were used as control. Error bars, SD ($n = 2$).

Table 1
List of reporter constructs useful for quantifying signal transduction in the *Arabidopsis* protoplast transient expression system (the reporters used in Fig. 2 are labeled with asterisks)

Name	Locus	GenBank accession	Use	ABRC[a] stock number	References
HBT-sGFP (S65T)	35S derivative	EF090408	Efficiency/ internal control	CD3-911	(17)
**RD29A-LUC*	At5g52310	EF090409	ABA/stress	CD3-912	(10, 18)
**AtGH3-LUC*	At2g23710	EF090410	Auxin	CD3-913	(10, 18)
WRKY29-LUC	At4g23550	EF090411	Innate immunity	CD3-914	(8)
GST6-LUC	At2g47730	EF090412	H_2O_2/ stress	CD3-915	(18)
HSP18.2-LUC	At5g59720	EF090413	Heat/H_2O_2	CD3-916	(10)
**ARR6-LUC*	At5g62690	EF090414	Cytokinin	CD3-917	(10)
GCC1-LUC	8× GCC-Box synthetic promoter	EF090415	Ethylene/ stress	CD3-918	(12)
**FRK1-LUC*	At2g19190	EF090416	Innate immunity	CD3-919	(9)

[a]*Arabidopsis* Biological Resource Center (http://abrc.osu.edu/index.html)

of isolation and high transfection efficiency make *Arabidopsis* mesophyll protoplasts a good system for transient expression of candidate genes to investigate their functions. We have successfully used luciferase (LUC) and β-glucuronidase (GUS) as reporters for quantitatively measuring the effect of regulators in different signal transduction pathways (8–14). A set of LUC reporters controlled by promoters that respond to different signals have been donated to the *Arabidopsis* Biological Resource Center (ABRC) for public use (Table 1). One drawback of the reporter assays is that same as other gain-of-function approaches transgene overexpression may cause an ectopic effect on the expression of reporter gene. However, the results of reporter assays can be validated by quantitative measurement of endogenous gene expression (e.g., marker genes for each signal pathway; Fig. 1) and further investigated in appropriate mutants. The assays can be applied for more ambitious high-throughput screening and systematic and genome-wide characterization of gene functions in most *Arabidopsis* accessions and mutant plants (8, 11, 13, 14).

2. Materials

2.1. Plant Material and Growth Conditions

Grow Arabidopsis plants in soil (Metro-Mix 360 or Jiffy-7) in a greenhouse or growth chamber with a photoperiod of 12 h light/12 h dark under low light (75 mmol/m^2/s) at a 23°C/20°C light/dark temperature regime and 65% relative humidity for 4 weeks.

2.2. Protoplast Isolation

1. Enzyme solution: 0.4 M mannitol, 20 mM KCl, 20 mM MES, pH 5.7, 1.5% cellulase R10 (Yakult Pharmaceutical Ind. Co., Ltd., Japan), 0.4% macerozyme R10 (Yakult Pharmaceutical Ind. Co., Ltd., Japan). Heat the enzyme solution at 55°C for 10 min to inactivate proteases and DNase and enhance enzyme solubility (see Note 1). Cool the solution to room temperature and add 10 mM $CaCl_2$, 1 mM β-mercaptoethanol, and 0.1% BSA. Pass the solution through a 0.45-μm nylon membrane syringe filter into a Petri dish.
2. Razor blades.
3. Petri dish (100 × 25 mm).
4. Desiccator.
5. Nylon mesh (35–75 μm).
6. 30-ml Round-bottom polypropylene tube (Sarstedt).
7. Hemacytometer.

2.3. Protoplast Transfection

1. W5 solution: 154 mM NaCl, 125 mM $CaCl_2$, 5 mM KCl, 2 mM MES, pH 5.7.
2. WI solution: 0.5 M mannitol, 4 mM MES, pH 5.7, 20 mM KCl.
3. MMg solution: 0.4 M mannitol, 15 mM $MgCl_2$, 4 mM MES, pH 5.7.
4. Polyethylene glycol (PEG) solution: Add 4 g of PEG 4000 (Sigma-Aldrich), 2.5 ml of 0.8 M mannitol, and 1 ml of 1 M $CaCl_2$ to 3 ml of ddH_2O for a total volume of 10 ml.
5. 2-ml Round-bottomed natural microcentrifuge tubes.
6. Tissue culture plates (6-, 12-, or 24-well).
7. Bench-top centrifuge (IEC Centra CL2; International Equipment Company).

8. Chemicals: 1-naphthaleneacetic acid (NAA; Sigma-Aldrich), (±)-abscisic acid (ABA; Sigma-Aldrich), 6-(γ,γ-dimethylallylamino)-purine (2-iP; Sigma-Aldrich), flg22 (the conserved 22 amino acids of flagellin, chemically synthesized according to the published peptide sequence (15)), and elf18 (N-terminal, acetylated 18 amino acid fragment of bacterial elongation factor Tu, synthesized according to the published peptide sequence (16)).

2.4. Quantitative Real-Time Reverse Transcription-Polymerase Chain Reaction Assay

1. TRIzol reagent (Invitrogen).
2. Oligo(dT)$_{15}$ primer (Promega).
3. dNTP (mix of dATP, dTTP, dGTP, and dCTP).
4. ImProm-II 5× Reaction Buffer (Promega).
5. 25 mM $MgCl_2$.
6. Protector RNase Inhibitor (Roche).
7. ImProm-II Reverse Transcriptase (Promega).
8. iQ SYBR Green Supermix (Bio-Rad).
9. CFX96 Real-Time polymerase chain reaction (PCR) Detection System (Bio-Rad).

2.5. Reporter Assays

1. Lysis buffer: 25 mM Tris-phosphate, pH 7.8, 2 mM 1, 2-diaminocyclohexane *N,N,N,N*-tetraacetic acid, 10% glycerol, 1% Triton X-100, 2 mM dithiothreitol (DTT).
2. Luciferase assay system (Promega).
3. MUG solution for GUS assay: 1 mM 4-methylumbelliferyl-β-D-glucuronide (MUG; Sigma-Aldrich, Cat. No. M9130), 10 mM Tris–HCl, pH 8, 2 mM $MgCl_2$.
4. Luminometer and fluorometer (Modulus microplate multimode reader; Promega).

2.6. In Vitro Kinase Assay

1. Immunoprecipitation (IP) buffer: 50 mM Tris–HCl, pH 7.5, 150 mM NaCl, 5 mM EDTA, 1 mM DTT, 2 mM NaF, 2 mM Na_3VO_4, 1× protease inhibitor cocktail (Roche), 1% Triton X-100.
2. Kinase buffer: 20 mM Tris–HCl, pH 7.5, 20 mM $MgCl_2$, 5 mM EGTA, 1 mM DTT, 50 μM ATP and 2 μCi ATP, [γ-^{32}P] or [γ-^{33}P] (PerkinElmer).
3. Protein G Sepharose beads (GE Healthcare).
4. Fixing solution: 10% ethanol and 10% acetic acid.
5. Gel dryer (Bio-Rad).
6. Typhoon imaging system (GE Healthcare).

3. Methods

3.1. Protoplast Isolation

1. Select well-expanded leaves from 3- to 4-week-old plants (see Note 2).
2. Remove the leaf tip (3 mm), cut the middle part of a leaf into 0.5- to 1-mm strips with a clean and sharp razor blade without crushing the edge, and immediately transfer the leaf strips into the solution and submerge the leaf strips (see Note 3).
3. Cover the Petri dish with aluminum foil and vacuum infiltrate for 30 min using a desiccator.
4. Continue the digestion without vacuum or shaking for another 2.5–3 h. The digestion time may vary depending on the material and experimental goals.
5. Release the protoplasts by gently shaking the Petri dish (either by hand or using a shaker at 50 rpm). Under the ideal conditions, the leaf strips turn transparent.
6. Add equal volume of W5 solution and filter the solution containing protoplasts with a 35- to 75-μm nylon mesh into a 30-ml round-bottom tube.
7. Pellet the protoplasts by spinning for 2 min at $100 \times g$ or speed 3 using an IEC clinical centrifuge.
8. Remove the supernatant as much as possible and resuspend the protoplasts in 0.5 ml of W5 solution by gentle swirling.
9. Count protoplasts using a hemacytometer under a light microscope and adjust the concentration to 2×10^5/ml by adding W5 solution.
10. Keep the protoplasts on ice for at least 30 min for recovery from isolation stress.
11. The protoplasts should settle at the bottom of the tube after 5–10 min. Right before PEG-Ca^{2+} transfection, pipette the W5 solution out and resuspend the protoplast pellet in MMg solution to 2×10^5/ml (see Note 4).

3.2. Protoplast PEG-Ca^{2+} Transfection

1. Prepare fresh 40% (*w/v*) PEG solution.
2. Add 10 μl (10–20 μg) of the plasmids DNA into a 2-ml round-bottom tube (see Note 5).
3. Add 100 μl of protoplasts in MMg solution into the tube (see Note 6).
4. Add 110 μl of PEG solution and then mix completely by gently tapping the tube.
5. Incubate at room temperature for up to 10 min (5 min is sufficient).

6. Add 440 μl of W5 solution and mix by gently inverting to stop the transfection.
7. Spin at 100 × *g* for 1 min and remove the supernatant.
8. Resuspend the protoplasts gently with 500 μl of WI solution in each well of a 12-well tissue culture plate (see Note 7).
9. Treat the protoplasts with plant hormones or microbe-associated molecular patterns (MAMPs) (optional; see Note 8).
10. Incubate the protoplasts under desirable conditions (see Note 9).
11. After incubation, resuspend and harvest protoplasts by centrifugation at 100 × *g* for 2 min.
12. Remove the supernatant and freeze the samples on dry ice. Samples can be stored at –80°C until further analysis.

3.3. Quantitative Real-Time Reverse Transcription-Polymerase Chain Reaction Assay

1. Isolate total RNA by using TRIzol Reagent according to the manufacturer's instructions.
2. Add 0.8 ml of TRIzol for 2×10^5 protoplasts (see Note 10).
3. Combine 1 μg of total RNA and 1 μl of Oligo$(dT)_{15}$ (500 ng/μl) in nuclease-free water for a final volume of 10 μl per RT reaction.
4. Heat the mixture at 70°C for 10 min and then chill on ice.
5. Add 10 μl of the reverse transcription reaction mix (4 μl of ImProm-II 5× reaction buffer, 2 μl of 25 mM $MgCl_2$, 2 mM dNTP, 0.5 μl of Protector RNase Inhibitor, and 0.5 μl of ImProm-II Reverse Transcriptase) to each reaction tube.
6. Incubate the tubes at 25°C for 5 min and then at 42°C for 1 h.
7. Inactivate the reverse transcriptase at 70°C for 15 min.
8. Add 80 μl of nuclease-free water to each RT product.
9. Take 1 μl of the diluted RT product for 10-μl quantitative real-time reverse transcription-PCR (qRT-PCR) with primers for genes of interest or control genes, such as "housekeeping" genes (e.g., Actin, Ubiquitin, or Tubulin genes). An example of the results produced is shown in Fig. 1.

3.4. Reporter Assays

3.4.1. Luciferase Activity Assay

1. Add 50 μl of cell lysis buffer to the frozen protoplast samples and thaw on ice (see Note 11).
2. Vortex vigorously for 10 s to break up the protoplasts and then incubate on ice for 5–10 min.
3. Spin down cell debris at 8,000–10,000 × *g* for 1 min.
4. Mix 2–20 μl of the lysate with 100 μl of LUC assay reagent to measure LUC activity with a luminometer (see Note 12).

3.4.2. GUS Activity Assay

1. Add 2 μl of the protoplast lysate prepared from Subheading 3.4.1, step 3, into 25 μl of MUG substrate solution and mix well.

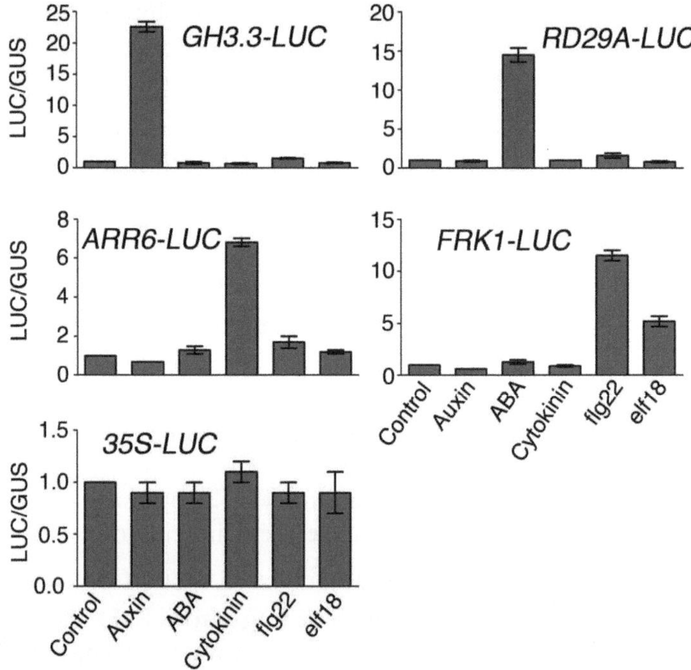

Fig. 2. Specificity of plant hormone or innate immune responses in the *Arabidopsis* protoplast transient expression system. Immediately after transfection, the protoplasts were treated with 10 μM NAA (auxin), 10 μM ABA, 100 nM 2-iP (cytokinin), 10 nM flg22, or 100 nM elf18. Protoplasts were harvested after 4 h of incubation. Promoter activities were normalized to the value obtained from protoplasts without treatment (control) and GUS activity of each sample served as an internal standard. Error bars, SD ($n = 2$).

2. Incubate at 37°C for 30–60 min.
3. Add 100 μl of 0.2 M Na_2CO_3 to stop the reaction.
4. Measure the fluorescence of MU using a fluorometer. Examples of the quantitative measurement of the reporter gene expression in response to specific signals are shown in Fig. 2.

3.5. In Vitro Kinase Assay

1. Add 200 μl of IP buffer to the frozen 200-μl (4×10^4) protoplast sample and thaw on ice.
2. Vortex vigorously for 10 s to break up the protoplasts and then incubate on ice for 5–10 min.
3. Centrifuge at the maximum speed for 5 min at 4°C.
4. Transfer the supernatant to a new tube and add 1 μl of antibody against the protein kinase of interest (see Note 13).
5. Incubate with constant mixing at 4°C for 2 h.
6. During the incubation time, wash Protein G Sepharose beads (5 μl per sample) with 1 ml of IP buffer.
7. Centrifuge at 6,000 rpm for 20 s and carefully remove the supernatant as much as possible.

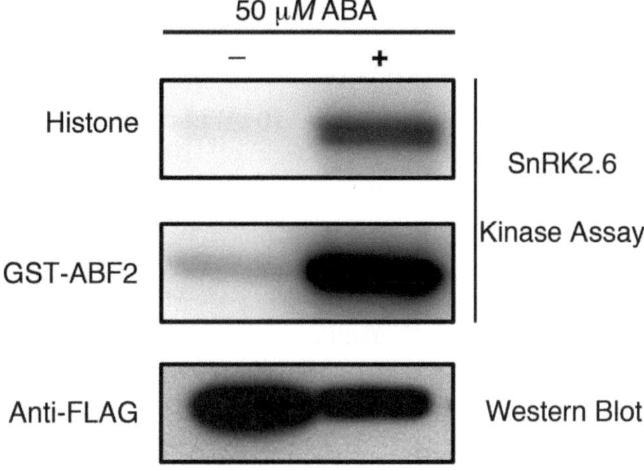

Fig. 3. Activation of transiently expressed SnRK2.6-FLAG by ABA in protoplasts. After transfection, the protoplasts were incubated in WI solution for 5.5 h and then treated with 50 μM ABA for 30 min. The control (–) was the duplicated protoplast sample incubated for 6 h without ABA. Two micrograms of histone and GST-ABF2 (Gly73 to Gln119) were used as substrates in kinase assays, and their phosphorylation by ABA-activated SnRK2.6 was detected by a phosphorimager (the *upper two panels*). The expression level of SnRK2.6-FLAG in protoplasts was detected by Western blot using anti-FLAG antibody (*bottom panel*).

8. Repeat steps 6 and 7 two more times.
9. After incubation with the antibody, add 5 μl of washed Protein G Sepharose beads into each sample and incubate at 4°C with constant mixing for another hour.
10. Centrifuge at 6,000 rpm for 20 s and carefully remove the supernatant.
11. Wash the beads with 1 ml of IP buffer twice.
12. Wash one more time with 1 ml of the kinase buffer (without cold and hot ATP).
13. Add 15 μl kinase buffer, including 1–3 μg of the desired substrate, into each sample and incubate at room temperature or 30°C for 30–60 min (see Note 14).
14. Stop the reaction by adding 4× SDS-PAGE loading buffer.
15. Boil the sample for 5 min and then load the supernatant of the reaction sample on a SDS-PAGE gel.
16. Run the gel at 30–50 mA until the dye fronts reach the bottom of the gel.
17. Wash the gel in the fixing buffer for 10 min three times.
18. Dry the gel on a gel dryer at 80°C for 1 h and then expose to autoradiography film or phosphorimager (Typhoon imaging system). ABA activation of SnRK2.6 is used as an example for kinase assay in Fig. 3.

4. Notes

1. Prepare 10 ml of the enzyme solution to digest up to 40 leaves with a yield of ~10^6 protoplasts per 10 leaves. After heating, the solution should be clear light brown. A workshop movie for protoplast isolation and transfection can be downloaded on the Sheen Lab Web site (http://genetics.mgh.harvard.edu/sheenweb/protocols_reg.html).
2. Younger plants can be used for protoplast isolation. Leaves five to eight are normally used in 4-week-old *Arabidopsis* plants. Leaves three to four are used in 3-week-old plants and leaves one to two are used in 2-week-old plants.
3. This is a critical step for protoplast isolation. Use fresh razor blades to cut the leaves on a piece of clean white paper to evaluate the process. There should not be juicy green stain left on the paper after cutting. In order to get a high yield of the protoplasts, completely submerge the leaf strips into the enzyme solution by gently dipping both sides with an inoculating loop (BD).
4. Be gentle with the protoplasts during all the steps! To resuspend the protoplast pellet, add 0.5–1 ml of the solution first and gently swirl the tube before adding the rest of the solution.
5. The quality of the plasmid DNA is critical for high transfection efficiency. We routinely use CsCl gradient for maxi-plasmid DNA preparation. The protocol is available on the Sheen Lab Web site (http://genetics.mgh.harvard.edu/sheenweb/protocols_reg.html).
6. Based on the purpose of the experiment, the experiments can be scaled up or down with the same DNA/protoplasts ratio. For the reporter assay, use 10–100 μl (0.2–2 × 10^4) protoplasts depending on the promoter activity. Use 100–200 μl (2–4 × 10^4) protoplasts for Western blot, coimmunoprecipitation, and kinase assay. For qRT-PCR analysis, 1 ml (2 × 10^5) protoplasts are normally required.
7. 24-Well (250 μl of WI), 6-well (1 ml of WI), or 100 × 25-mm Petri dish (5 ml of WI) can also be used depending on the amount of protoplasts (~4 × 10^4 cells/ml of WI). The plate is coated with 5% sterile calf serum for 1 s to prevent the protoplast to stick on the surface. The depth of the WI solution in the plate is about 0.1 mm in order to prevent hypoxia stress during incubation.
8. Treat the protoplasts immediately if carrying out reporter assays in order to obtain a low basal level of reporter gene expression. To measure the expression level of endogenous genes in

response to a stimulus, it is highly recommended to rest the protoplasts in the WI solution for 3–5 h before signal treatment.

9. The incubation conditions can vary depending on the experiment purposes. We usually incubate the protoplasts at room temperature (22–25°C) under low light (30–35 μmol/m^2/s). Normally, 3–6 h is enough for reporter assay and Western blot. For qRT-PCR analysis, different time points after treatment can be chosen based on the experimental plan. The kinase activity can be detected within minutes to 1 h after treatment.
10. The total RNA yield is about 5 μg for 2×10^5 protoplasts.
11. Add fresh DTT to the cell lysis buffer right before use.
12. The amount of the lysate used in LUC assay depends on the promoter activity and the expression level of the reporter. Make sure that the readings of samples are within the linear range. The LUC assay reagent (Promega, Cat. No. E1501) can be diluted five times with water right before use.
13. The protoplast system is an ideal system for coimmunoprecipitation or kinase assays when the antibody against the endogenous protein is not available. Add an epitope tag sequence (e.g., HA, MYC, and FLAG tag) onto the gene of interest and express it in the protoplasts. Then, use the anti-eptitope antibody to immunoprecipitate the protein of interest.
14. The composition of the kinase buffer is specific to the kinase of interest. The kinase buffer used in this protocol is suitable for MAPK and SnRK2.6 kinase assay.

Acknowledgments

We thank the present and past members of the Sheen Laboratory for their efforts to set up and improve the Arabidopsis mesophyll protoplasts transient expression system. We gratefully acknowledge Dr. Guillaume Tena, Dr. Kun-Hsiang Liu, Dr. Yan Xiong, Dr. Joonyup Kim, and Dr. Matthew Ramon for their technical help. We also would like to thank our greenhouse manager Jenifer Bush for providing plant material. This work was supported by NSF IOS 0618292, NIH R01 GM070567, and MGH-CCIB fund.

References

1. Sheen J (2001) Signal transduction in maize and Arabidopsis mesophyll protoplasts. Plant Physiol 127:1466–1475
2. Browse J (2009) The power of mutants for investigating jasmonate biosynthesis and signaling. Phytochemistry 70:1539–1546

3. Liscum E, Reed JW (2002) Genetics of Aux/IAA and ARF action in plant growth and development. Plant Mol Biol 49:387–400
4. Park SY et al (2009) Abscisic acid inhibits type 2 C protein phosphatases via the PYR/PYL family of START proteins. Science 324:1068–1071
5. Cocking EC (1960) A Method for the isolation of plant protoplasts and vacuoles. Nature 187:927–929
6. Fujii H et al (2009) In vitro reconstitution of an abscisic acid signalling pathway. Nature 462:660–664
7. Tena G et al (2001) Plant mitogen-activated protein kinase signaling cascades. Curr Opin Plant Biol 4:392–400
8. Asai T et al (2002) MAP kinase signalling cascade in Arabidopsis innate immunity. Nature 415:977–983
9. He P et al (2006) Specific bacterial suppressors of MAMP signaling upstream of MAPKKK in Arabidopsis innate immunity. Cell 125:563–575
10. Hwang I, Sheen J (2001) Two-component circuitry in Arabidopsis cytokinin signal transduction. Nature 413:383–389
11. Muller B, Sheen J (2008) Cytokinin and auxin interaction in root stem-cell specification during early embryogenesis. Nature 453:1094–1097
12. Yanagisawa S, Yoo SD, Sheen J (2003) Differential regulation of EIN3 stability by glucose and ethylene signalling in plants. Nature 425:521–525
13. Baena-Gonzalez E et al (2007) A central integrator of transcription networks in plant stress and energy signaling. Nature 448:938–942
14. Boudsocq M et al (2010) Differential innate immune signalling via Ca(2+) sensor protein kinases. Nature 464:418–422
15. Felix G et al (1999) Plants have a sensitive perception system for the most conserved domain of bacterial flagellin. Plant J 18:265–276
16. Kunze G et al (2004) The N terminus of bacterial elongation factor Tu elicits innate immunity in Arabidopsis plants. Plant Cell 16:3496–3507
17. Chiu W et al (1996) Engineered GFP as a vital reporter in plants. Curr Biol 6:325–330
18. Kovtun Y et al (2000) Functional analysis of oxidative stress-activated mitogen-activated protein kinase cascade in plants. Proc Natl Acad Sci 97:2940–2945

Chapter 17

Genome-Wide Profiling of Uncapped mRNA

Yuling Jiao and José Luis Riechmann

Abstract

Gene transcripts are under extensive posttranscriptional regulation, including the regulation of their stability. A major route for mRNA degradation produces uncapped mRNAs, which can be generated by decapping enzymes, endonucleases, and small RNAs. Profiling uncapped mRNA molecules is important for the understanding of the transcriptome, whose composition is determined by a balance between mRNA synthesis and degradation. In this chapter, we describe a method to profile these uncapped mRNAs at the genome scale.

Key words: mRNA, Uncapping, RNA degradation, Transcriptome

1. Introduction

Widely used high-throughput transcriptome profiling approaches have been successful in dissecting gene regulatory networks. Still, transcriptome-wide profiling is often compromised by the complicated life cycle of mRNA, from transcription to decay. The different forms of RNA molecules that exist throughout the mRNA life cycle (from full-length, capped RNA to uncapped or cleaved RNA) are often indistinguishable by traditional RNA profiling. As the abundance of mRNA within cells is determined by the rates of mRNA synthesis and degradation, the reconstruction of gene expression networks clearly requires data for mRNA degradation and other modes of regulation of mRNA transcript abundance. Numerous studies indicate that mRNA degradation is a determining factor for the steady-state levels of mRNAs in cells. The decay of mRNA, in turn, can be affected by various developmental and environmental stimuli (1, 2).

Zhi-Yong Wang and Zhenbiao Yang (eds.), *Plant Signalling Networks: Methods and Protocols*,
Methods in Molecular Biology, vol. 876, DOI 10.1007/978-1-61779-809-2_17,

The degradation of mRNA in eukaryotes can be initiated by one of several highly conserved pathways. Usually, general mRNA decay is initiated by deadenylation via a variety of mRNA deadenylases that shorten the 3′ poly(A) tail (3). A decapping enzyme complex consisting of DCP1 and DCP2 then removes the 5′-modified guanine nucleotide cap structure. The decapped transcripts are progressively digested by a 5′–3′ exonuclease known as XRN1 in yeast and human (2). As an alternative, deadenylated mRNAs may be degraded in a 3′–5′ direction by the cytoplasmic exosome complex. In addition, nonsense-mediated mRNA decay (NMD) is a quality control system that rapidly removes mRNAs containing premature termination codons through deadenylation-independent decapping. Endonuclease cleavage can also initiate mRNA degradation, mediated either indirectly by small RNA-mediated silencing or directly by endonuclease-mediated cleavage, both of which generate uncapped mRNA. These pathways (with the exception of the 3′–5′ decay pathway) produce mRNA fragments with a free 5′ monophosphate group.

The presence of this free 5′ phosphate has been exploited to design the RNA ligase-mediated 5′-RACE to map cleavage sites of individual transcripts (4, 5). More recently, the method has been extended to the genome scale to map all uncapped mRNA molecules (6–10). Briefly, the free 5′ phosphate group of uncapped mRNA is used for the T4 RNA ligase-mediated ligation to an RNA adaptor, which is subsequently used for RNA purification and selective cDNA synthesis. The resulting cDNA can then be profiled by microarray hybridizations or high-throughput sequencing (Fig. 1).

2. Materials

1. Nuclease-free sterile water.
2. Nuclease-free sterile tubes.
3. Total RNA isolation kit or reagents, such as the RNeasy mini kit (Cat. No. 74104, Qiagen, Valencia, CA).
4. Poly(A)$^+$ RNA purification kit, such as the Micro-FastTrack 2.0 kit (Cat. No. K1520-02, Invitrogen, Carlsbad, CA).
5. RNA Adaptor: 5′-CGA CUG GAG CAC GAG GAC ACU GAC AUG GAC UGA AGG AGU AGA AA-3′.
6. T4 RNA ligase (5 U/μl) and 10× ligase buffer (Ambion).
7. Ribonuclease inhibitor, such as RNasin (40 U/μl) (Promega).
8. 3′ Biotinylated DNA probe: 5′-GTC CTC GTG CTC CAG TCG/3BioTEG/-3′.

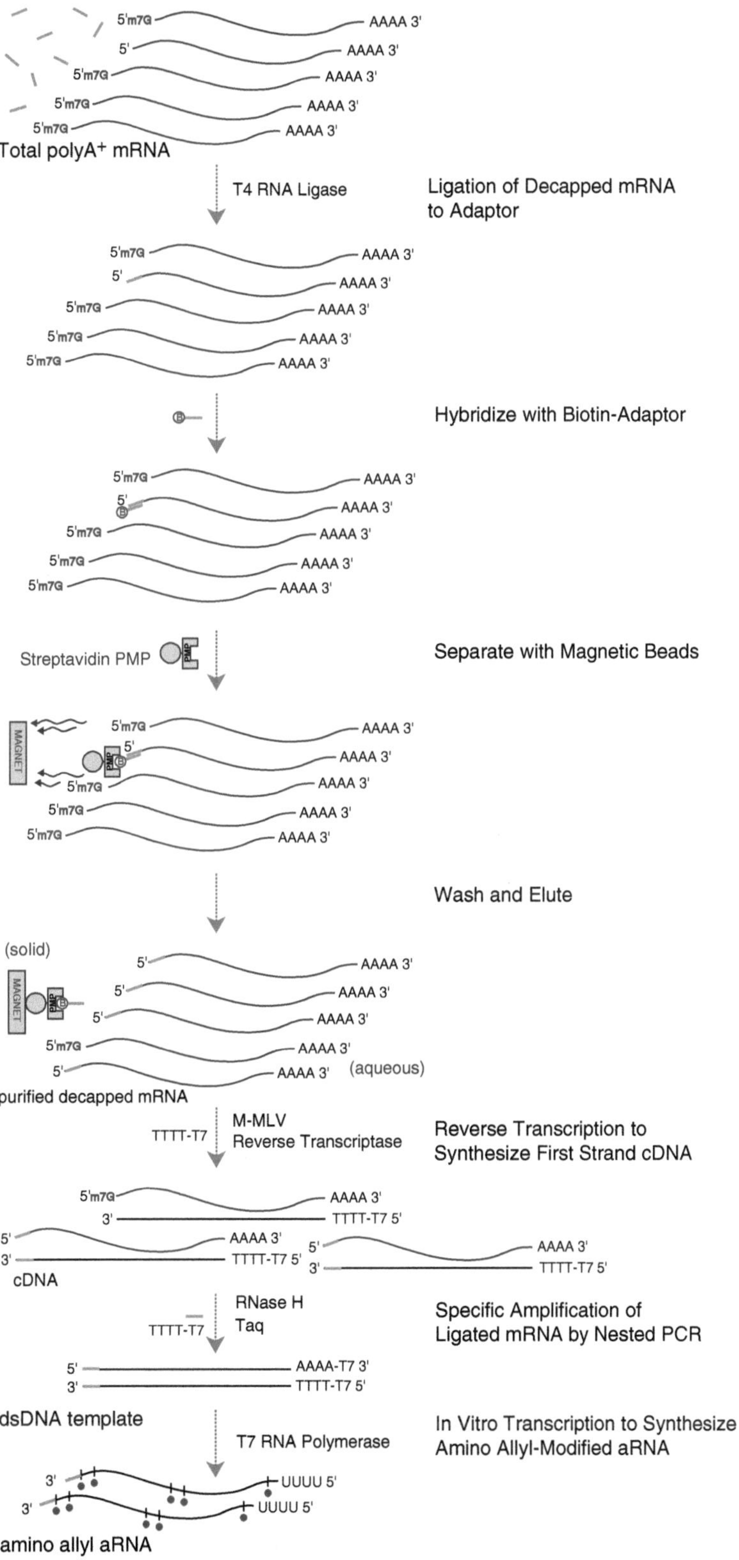

Fig. 1. Schematic description of the steps for isolation of uncapped mRNA, reverse transcription, and labeled aRNA amplification for microarray hybridization.

9. Streptavidin-paramagnetic particles (SA-PMPs) (Promega).
10. T7dT primer: 5′-GGC CAG TGA ATT GTA ATA CGA CTC ACT ATA GGG AGG CGG TTT TTT TTT TTT TTT TTT TTT TTT-3′.
11. Superscript II RT (200 U/μl) (Invitrogen).
12. dNTPs, 2.5 mM each.
13. DTT, 0.1 M.
14. RNase H (2 U/μl) (Cat. No. 18021-014, Invitrogen).
15. Taq DNA polymerase, such as ExTaq (5 U/μl) (Cat. No. TAK RR001A, Clontech, Madison, WI).
16. GR primer: 5′-CGA CTG GAG CAC GAG GAC ACT GA-3′.
17. GRN primer: 5′-GGA CAC TGA CAT GGA CTG AAG GAG TA-3′.
18. PCR cleanup kit, such as DNA Clean and Concentrator (Cat. No. D4003, Zymo, Orange, CA).
19. T7 RNA polymerase, such as the Megascript T7 kit (Ambion).
20. ATP, CTP, and GTP mix, 25 mM each.
21. UTP, 50 mM.
22. Amino allyl-UTP (aa-UTP), 50 mM (Ambion).
23. 0.4 M Na_2CO_3, pH 8.5.
24. Cy3 and Cy5 mono-reactive dye (GE Healthcare).
25. 4 M hydroxylamine.
26. 10× Fragmentation reagent and stop solution (Ambion).
27. Spin-50 mini-columns.
28. T7 Oligo(dT) primer: 5′-GGC CAG TGA ATT GTA ATA CGA CTC ACT ATA GGG AGG CGG TTT TTT TTT TTT TTT TTT TTT TTT-3′.

3. Methods

3.1. Extract Poly(A)$^+$ RNA

1. Isolate total RNA using the RNeasy kit, or a similar method, according to the manufacturer's instructions. At least 500 μg of total RNA is required per experiment.
2. Isolate poly(A)$^+$ RNA from the total RNA using the Micro-FastTrack 2.0 kit according to the manufacturer's instructions. This method generally yields >500 ng of poly (A)$^+$ RNA.

3.2. Adaptor Ligation

1. Vacuum dry 500 ng of poly (A)$^+$ RNA to 1 μl.
2. Assemble the adaptor ligation mix in a nuclease-free tube on ice:

Poly $(A)^+$ RNA	500 ng
RNA adaptor (0.3 µg/µl)	1 µl
10× T4 RNA ligase buffer	1 µl
T4 RNA ligase (5 U/µl)	2 µl
RNasin (40 U/µl)	1 µl

Nuclease-free water to a final volume of 10 µl.

3. Mix well and incubate reactions for 1.5 h at 37°C.

3.3. Uncapped Poly $(A)^+$ RNA Purification

1. Add 240 µl of nuclease-free water to the ligation reaction.
2. Place the tube in a 65°C heating block for 10 min.
3. Add 1.5 µl of the (50 pM/µl) 3′ biotinylated DNA probe and 6.5 µl of 20× SSC. Mix gently and incubate at room temperature until completely cooled (~10 min or less).
4. Prepare 1.2 ml of sterile 0.5× SSC by combining 30 µl of 20× SSC with 1.17 ml of nuclease-free water.
5. Prepare 1.4 ml of sterile 0.1× SSC by combining 7 µl of 20× SSC with 1.393 ml of nuclease-free water.
6. Resuspend one tube (0.6 ml) of the SA-PMPs per two isolations by gently flicking the bottom of the tube until the particles are completely dispersed, and then capture them by placing the tube in the magnetic stand until the SA-PMPs have been collected at the side of the tube (~30 s).
7. Carefully remove the supernatant while the tube sits in the magnetic stand.
8. Wash the SA-PMPs three times with 0.5× SSC (300 µl per wash), each time capturing them using the magnetic stand and carefully removing the supernatant.
9. Resuspend the washed SA-PMPs in 100 µl of 0.5× SSC and split into two tubes. Use one tube per reaction.
10. Add the entire contents of the annealing reaction (from step 3) to one tube (~50 µl) of washed SA-PMPs.
11. Incubate at room temperature for 10 min. Gently mix by inverting every 1–2 min.
12. Capture the SA-PMPs using the magnetic stand and carefully remove the supernatant without disturbing the SA-PMP pellet.
13. Wash the particles four times with 0.1× SSC (150 µl per wash) by gently flicking the bottom of the tube until all the particles are resuspended. After the final wash, remove as much of the supernatant as possible without disturbing the SA-PMPs.

14. Elute the mRNA by resuspending the final SA-PMP pellet in 50 μl of nuclease-free water (preheated to 55°C). Gently resuspend the particles by flicking the tube.
15. Magnetically capture the SA-PMPs and transfer the eluted RNA to a sterile, nuclease-free tube. Do not discard the particles.
16. Repeat the elution step by resuspending the SA-PMP pellet in 75 μl of nuclease-free water (preheated to 55°C). Repeat the capture step, pooling the eluate with the RNA eluted in the previous step.
17. Freeze the eluate at −20°C for 10 min, and then vacuum dry to reduce the volume to 7 μl.

3.4. Reverse Transcription to Synthesize First-Strand cDNA

1. Add 1 μl (100 ng/μl) of T7dT primer.
2. Incubate for 10 min at 70°C in a heating block. Centrifuge briefly.
3. Assemble the reverse transcription mix in a nuclease-free tube at RT:

5× First-strand buffer	4 μl
0.1 M DTT	2 μl
2.5 mM dNTP mix	4 μl
RNasin (40 U/μl)	1 μl
SuperScript II RT (200 U/μl)	1 μl

4. Mix well and transfer 12 μl of the mix to each RNA sample.
5. Incubate reactions for 2 h at 42°C.

3.5. Second-Strand cDNA Synthesis

1. Add 1 μl of RNase H (2 U/μl) and incubate for 15 min at 37°C, and then transfer to ice.
2. Assemble the first PCR mix in a nuclease-free tube on ice:

10× ExTaq buffer	18 μl
2.5 mM each dNTP mix	16 μl
20 μM GR primer	8 μl
T7dT primer (100 ng/μl)	8 μl
ExTaq (5 U/μl)	2 μl
Nuclease-free water	127 μl

3. Mix well and transfer 179 μl of the mix to each sample on ice. Mix well and split into four PCR tubes on ice.

4. The PCR program is (see Note 1) as follows:
   (a) 94°C for 2 min.
   (b) 94°C for 30 s.
       72°C for 3 min.
       Repeat (b) for two cycles.
   (c) 94°C for 30 s.
       70°C for 3 min.
       Repeat (c) for two cycles.
   (d) 94°C for 30 s.
       66°C for 30 s.
       68°C for 3 min.
       Repeat (d) for two cycles.
   (e) 72°C for 10 min.
5. Combine the tubes and purify with DNA Clean and Concentrator (Zymo), or a similar kit, following the manufacturer's instructions. Elute with nuclease-free water into 16 μl.
6. Assemble the second PCR mix in a nuclease-free tube on ice:

10× ExTaq buffer	5 μl
2.5 mM dNTP mix	4 μl
20 μM GRN primer	1 μl
T7dT primer (100 ng/μl)	1 μl
ExTaq (5 U/μl)	0.5 μl
Nuclease-free water	22.5 μl

7. Mix well and transfer 34 μl of the mix to each sample in a PCR tube on ice.
8. The PCR program is as follows:
   (a) 94°C for 2 min.
   (b) 94°C for 30 s.
       65°C for 30 s.
       72°C for 3 min.
       Repeat (b) for five cycles.
   (c) 72°C for 10 min.
9. Purify the PCR products with DNA Clean and Concentrator (Zymo), or a similar kit, following the manufacturer's instructions. Elute with nuclease-free water into 14 μl (see Note 2).

3.6. In Vitro Transcription to Synthesize aRNA

1. Assemble the transcription mix in a nuclease-free tube at room temperature:

10× Reaction buffer	4 μl
25 mM ATP, CTP, and GTP mix	12 μl
50 mM UTP	3 μl
50 mM aa-UTP	3 μl
T7 enzyme mix	4 μl

2. Mix well and transfer 26 μl of the mix to each sample. Mix thoroughly by pipetting up and down two to three times, and then flicking the tube three to four times, and centrifuge briefly.
3. Incubate reactions for ~14 h at 37°C in a hybridization oven.
4. Stop the reaction by adding 60 μl nuclease-free water.
5. Clean up the RNA using the RNeasy kit, or a similar kit, following the manufacturer's instructions. Elute with nuclease-free water into 100 μl.
6. Determine the concentration of each aRNA sample by measuring its absorbance at 260 nm using a spectrophotometer, such as NanoDrop 1000A (see Note 3).

3.7. Dye Coupling

1. Resuspend one Cy3 or Cy5 dye vial in 88 μl DMSO (see Note 4).
2. Place 5–20 μg of aRNA in an amber microcentrifuge tube and vacuum dry on medium or low heat until no liquid remains (see Note 5).
3. Add 7 μl of nuclease-free water and 2 μl of 0.4 M Na_2CO_3 (pH 8.5) to the dried aRNA and resuspend thoroughly by vortexing gently.
4. Add 11 μl of the corresponding dye, previously resuspended in DMSO. Mix well by vortexing gently.
5. Incubate the reaction in the dark at room temperature for 30 min.
6. Add 4.5 μl of 4 M hydroxylamine and mix well by vortexing gently.
7. Incubate the reaction in the dark at room temperature for 15 min.
8. Add 5.5 μl of nuclease-free water and bring the volume to 30 μl.
9. Clean up the labeled aRNA using RNeasy kit (Qiagen), or a similar kit, following the manufacturer's instructions. Elute with nuclease-free water into 30 μl.

10. Vacuum dry the labeled aRNA to 9 μl.
11. Add 1 μl of 10× fragmentation reagent and mix well.
12. Incubate the reaction at 70°C in a heating block for 10 min.
13. Add 1 μl of stop solution and 15 μl of nuclease-free water.
14. Prepare a Spin-50 column by spinning for 3 min at 1,000 × *g*. Transfer the column to a nuclease-free amber tube.
15. Load the entire reaction from step 12 onto the center of the Spin-50 column, and then centrifuge for 3 min at 1,000 × *g*.
16. Determine the concentration and dye incorporation rate of the aRNA sample by measuring its absorbance at 260, 550, and 650 nm using a NanoDrop 1000A spectrophotometer (see Note 6).

4. Notes

1. The indicated PCR cycling parameters are optimized for using ExTaq DNA polymerase with an MJ Research PTC-200 thermo cycler and with *Arabidopsis thaliana* RNA. The annealing temperatures may need further optimization if other conditions are used. To obtain the most suitable parameters, it is recommended to extend the PCR reaction with additional cycles and to analyze by agarose gel electrophoresis the PCR products that are obtained with different annealing temperatures.
2. Although the following steps produce labeled aRNA for microarray hybridization, direct sequencing may also be used to quantify uncapped poly $(A)^+$ RNA and to map the 5′ ends.
3. The size distribution of aRNA can be evaluated using an Agilent 2100 bioanalyzer or by denaturing agarose gel electrophoresis. The expected profile of the aRNA sample is a distribution of sizes from 250 to 5,000 nt, with the smear peaking at around 1,000–1,500 nt.
4. Cy3 and Cy5 reactive dyes are sensitive to light and water. Unused DMSO-dissolved dye should be stored at −20°C for up to 1 month and be out of moisture.
5. Check the progress of drying every 5–10 min, and do not overdry.
6. The dye incorporation rate, i.e., the number of dye molecules incorporated per 1,000 nt of labeled aRNA, can be estimated using the following formula:

$$\text{Incorporation rate} = (A_{\text{dye}}/A_{260}) \times (9{,}010/\text{cm}/\text{M})/ (\text{dye extinction coefficient}) \times 1{,}000.$$

Cy3 has an absorbance maximum at 550 nm and a dye extinction coefficient of 150,000. Cy5 has an absorbance maximum at 650 nm and a dye extinction coefficient of 250,000.

Acknowledgments

The authors would like to thank Dr. E.M. Meyerowitz for his encouragement and advice. This work was supported by the US National Science Foundation 2010 Project Grant 0520193, by the National Natural Science Foundation of China Grant 31171159, and by the Millard and Muriel Jacobs Genetics and Genomics Laboratory at the California Institute of Technology.

References

1. Gutierrez RA, MacIntosh GC, Green PJ (1999) Current perspectives on mRNA stability in plants: multiple levels and mechanisms of control. Trends Plant Sci 4:429–438
2. Parker R, Song H (2004) The enzymes and control of eukaryotic mRNA turnover. Nat Struct Mol Biol 11:121–127
3. Mitchell P, Tollervey D (2000) mRNA stability in eukaryotes. Curr Opin Genet Dev 10:193–198
4. Liu X, Gorovsky MA (1993) Mapping the 5′ and 3′ ends of tetrahymena thermophila mRNAs using RNA ligase mediated amplification of cDNA ends (RLM-RACE). Nucleic Acids Res 21:4954–4960
5. Llave C, Xie Z, Kasschau KD, Carrington JC (2002) Cleavage of scarecrow-like mRNA targets directed by a class of Arabidopsis miRNA. Science 297:2053–2056
6. Addo-Quaye C, Eshoo TW, Bartel DP, Axtell MJ (2008) Endogenous siRNA and miRNA targets identified by sequencing of the Arabidopsis degradome. Curr Biol 18:758–762
7. German MA, Pillay M, Jeong DH, Hetawal A, Luo S, Janardhanan P, Kannan V, Rymarquis LA, Nobuta K, German R, De Paoli E, Lu C, Schroth G, Meyers BC, Green PJ (2008) Global identification of microRNA-target RNA pairs by parallel analysis of RNA ends. Nat Biotechnol 26:941–946
8. Gregory BD, O'Malley RC, Lister R, Urich MA, Tonti-Filippini J, Chen H, Millar AH, Ecker JR (2008) A link between RNA metabolism and silencing affecting Arabidopsis development. Dev Cell 14:854–866
9. Jiao Y, Riechmann JL, Meyerowitz EM (2008) Transcriptome-wide analysis of uncapped mRNAs in Arabidopsis reveals regulation of mRNA degradation. Plant Cell 20:2571–2585
10. Franco-Zorrilla JM, Del Toro FJ, Godoy M, Perez-Perez J, Lopez-Vidriero I, Oliveros JC, Garcia-Casado G, Llave C, Solano R (2009) Genome-wide identification of small RNA targets based on target enrichment and microarray hybridizations. Plant J 59:840–850

Chapter 18

Computational Tools for Quantitative Analysis of Cell Growth Patterns and Morphogenesis in Actively Developing Plant Stem Cell Niches

Anirban Chakraborty, Ram Kishor Yadav, Min Liu, Moses Tataw, Katya Mkrtchyan, Amit Roy Chowdhury, and G. Venugopala Reddy

Abstract

Pattern formation in developmental fields involves precise spatial arrangement of different cell types in a dynamic landscape wherein cells exhibit a variety of behaviors, such as cell division, cell expansion, and cell migration [Reddy (Curr Opin Plant Biol 11:88–931, 2008) and Meyerowitz (Cell 88:299–3082, 2007)]. The information is exchanged between multiple cell layers through cell–cell communication processes to regulate gene expression and cell behaviors in specifying distinct cell types. Therefore, a quantitative and dynamic understanding of the spatial and temporal organization of gene expression and cell behavioral patterns within multilayered and actively growing developmental fields is crucial to model the process of development. The quantification of spatiotemporal dynamics of cell behaviors requires computational tools in image analysis, statistical modeling, pattern recognition, machine learning, and dynamical system identification. Here, we give a brief account of recently developed methods in analyzing both local and global growth patterns in *Arabidopsis* shoot apical meristems. The computational toolkit can be used to gain new insights into causal relationships among cell growth, cell division, changes in gene expression patterns, and organ development by analyzing various mutants that affect these processes. This may allow us to develop function space models that capture variations in several growth parameters both at local/single-cell level and at global/organ level. In the long run, this may enable clustering of molecular pathways that mediate distinct cell behaviors.

Key words: Arabidopsis, Shoot apical meristem, Cell segmentation, Cell tracking, Self-renewal, Differentiation, Function space model

1. Introduction

1.1. Shoot Apical Meristem Stem Cell Niche

Stem cell maintenance in the shoot apical meristems (SAMs) is a process in which the balance between self-renewal and differentiation of stem cell progeny is dynamically regulated through cell–cell communication networks (1). The SAM of *Arabidopsis thaliana*

Zhi-Yong Wang and Zhenbiao Yang (eds.), *Plant Signalling Networks: Methods and Protocols*,
Methods in Molecular Biology, vol. 876, DOI 10.1007/978-1-61779-809-2_18,

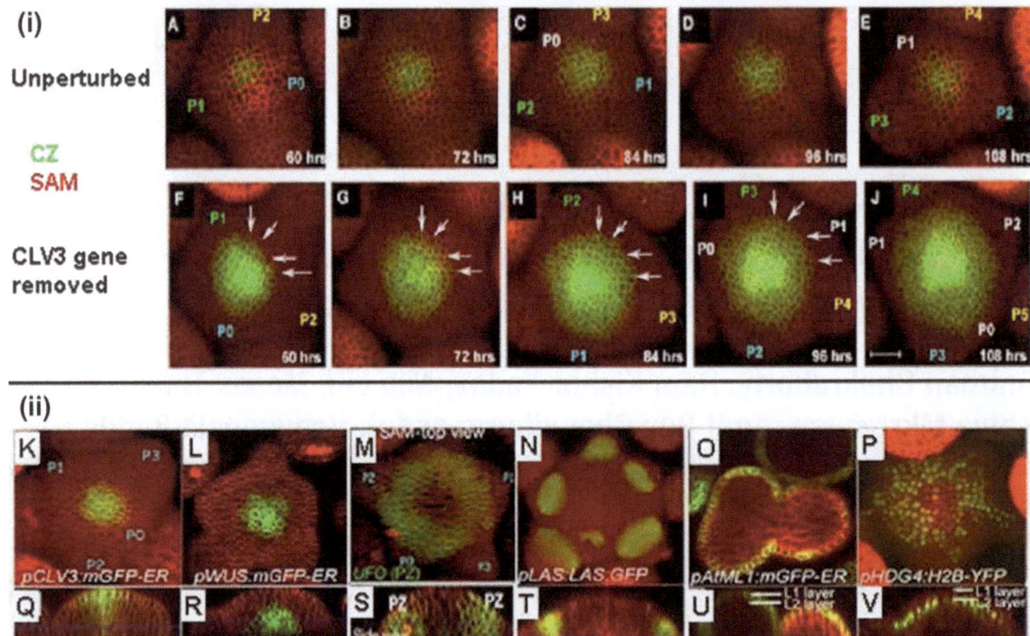

Fig. 1. **(i)** Local growth patterns: time-lapse series showing the sequential expansion of the stem cell domain (CZ) upon *CLV3* silencing (*F–J*). (*A–E, F–J*) Reconstructed 3D views of the shoot apical meristems (SAMs) in which the CLV3 gene is not perturbed and CLV3 gene has been transiently removed, respectively. Time elapsed after gene perturbation is marked on each panel. The CZ behavior (*pCLV3::mGFP5-ER, green*) is followed along with a cell division marker (*35S::YFP29-1, red*). The sequential expansion of the stem cell domain is due to the respecification of differentiating peripheral zone (PZ) cells into stem cells (*F–J*). **(ii)** Global growth patterns: Development of cell type-specific gene expression map. (*K–P*) 3D reconstructed views of SAMs expressing various GFP reporter lines highlighting the stem cell domain (CZ) marked by *CLAVATA3* promoter (*K*); the adjacent peripheral zone (PZ) is marked by *UFO* promoter (*M*); *A* section of the Rib-meristem (RM) revealing *WUSCHEL* expression (*L*); and the sites of differentiation are marked by *LAS* promoter (*N*), L1 layer-specific promoter (*O*), and L2 layer-specific promoter (*P*). (*Q–V*) Reconstructed side views of expression patterns shown in (*K*)–(*P*). Cell outlines are highlighted by FM4-64 dye (*red*).

consists of about 500 cells, located at the tip of the shoot, and it gives rise to all of the aboveground organs. The central zone (CZ) of the SAM harbors a set of stem cells (Fig. 1K, Q). The progeny of stem cells enter the flanking peripheral zone (PZ) (Fig. 1M, N, S, T) and the Rib meristem (RM) (Fig. 1L, R), where they differentiate. The cells of the RM function as niche cells by providing signals for maintenance of stem cells in the CZ. Apart from this radial organization, the SAM of *Arabidopsis* is a multilayered structure in which cells are organized into three clonally distinct layers (Fig. 1L, R). The cells in the outermost L1 layer (Fig. 1O, U) and the subepidermal L2 layer (Fig. 1P, V) divide in anticlinal orientation (perpendicular to the SAM surface), while the underlying corpus forms a multilayered structure where cells divide in periclinal (parallel to the SAM surface) as well as anticlinal (perpendicular to the SAM surface) planes (2, 3). Thus, the SAM stem cell niche represents a dynamic 3D network in which cell growth dynamics, gene expression, and organ growth are dynamically regulated so that the correct cellular identity, shape, and size of the SAM are preserved. The challenge is to understand how the interconnected network of cells interprets complex spatiotemporal signals to regulate gene expression patterns and cell behaviors.

1.2. Cell–Cell Communication

To achieve the above, it is essential to identify what genes are involved in these processes and how they regulate individual cell behaviors, such as cell division rates, cell expansion patterns, timing of cell division, and timely transition of cells from one gene expression state to another. Both *bottom-up* and *top-down* approaches have been employed to identify genes that control cell behaviors and gene expression patterns within the SAM stem cell niche (4).

1.2.1. Bottom-Up Approaches

Earlier studies, based on genetic and molecular analysis, have resulted in a molecular framework of cell–cell communication machinery involved in stem cell homeostasis (4). However, most of our current understanding comes from single-time-point observations that do not take into account the dynamics of cell behavior and gene expression. Cell behavior is tightly coupled, both in space and time, to cell identity transitions such that the effects they exert on each other cannot be easily discerned from such static studies. Recent studies have employed novel microscopic methods to obtain real-time observations of cell behaviors and gene expression patterns and they have also been coupled to transient manipulation of gene activities (5–7). These studies have revealed an accurate spatiotemporal dynamics of the influence of gene activities on cell behaviors and gene expression.

The novel live-imaging method utilizes laser scanning confocal microscopy to acquire time series of the 3D imagery of SAMs for up to 4–5 days. An example of one such 4D data set in which the SAM is labeled with plasma membrane-localized yellow fluorescent protein (YFP) to follow individual cell divisions is shown in Fig. 2i. The manual tracking of cells through successive cell divisions and gene expression patterns is beginning to yield new insights into the process of stem cell homeostasis (8). However, manual tracking is laborious and impossible as larger and larger amounts of microscope imagery are collected worldwide. Also a mere manual analysis of these data sets may not reveal quantitative information of cell behavior dynamics, and therefore a large part of this data will be underutilized not only in SAMs but also in several animal and plant developmental fields, where such live-imaging methods are being employed. Therefore, we require tools for cell tracking, lineage computation, and shape modeling for analyzing large volumes of data.

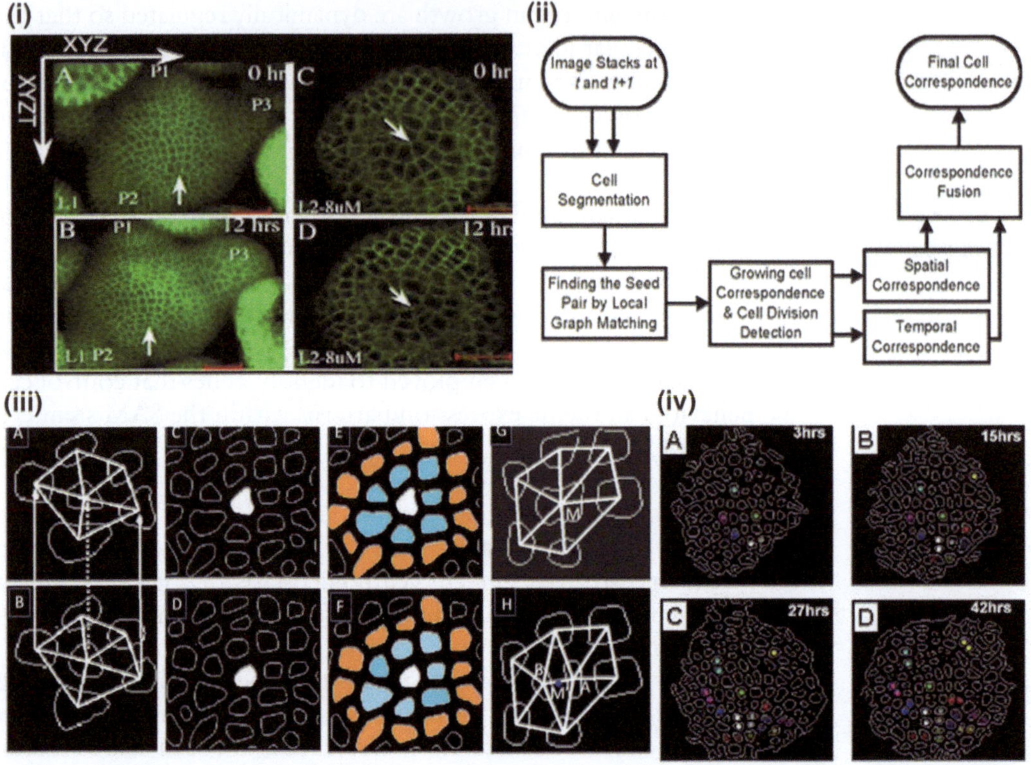

Fig. 2. (**i**) 4D Fluorescent imagery of shoot apical meristem stem cell niche. (*A–D*) 35S::YFP29-1 (plasma membrane-localized yellow fluorescent protein) expression in different layers of cells in the SAMs. (*A*, *C*) First time points (0 h) and (*B*, *D*) the same SAMs after the elapsed time indicated on top right-hand corner of each panel. (*A*, *B*) 3D-reconstructed Z-stack revealing all the cells in the L1 layer. Differentiating organs at different stages of development are marked as P1, P2, and P3. *Arrows* indicate the same cells prior to and after division. (*C*, *D*) Cells in the L2 layer (8 μm deep) and the *arrow* points to cell division events. Scale bar 20 μm. *XYZ* represents 3D image as stacks. *XYZT* represents 3D image across time. (**ii**) Block diagram of cell lineage computation framework using our robust tracking method (19). (**iii**) (*A*, *B*) Local graph matching. (*C*, *D*) Seed pair detection. (*E*, *F*) Growing correspondence from the seed pair. (*G*, *H*) Cell division detection. (**iv**) Lineages of some sample cells (*different colors*) over many hours of imaging. Total time elapsed is indicated on each panel.

1.2.2. Top-Down Approaches

Though "bottom-up" approaches provide information on a few crucial molecules within the network, it may not reveal all molecules that are part of an integrated network. However, the advances in genome biology in the past decade have resulted in large-scale description of gene expression patterns in developmental fields. Expression profiling using microarrays provides only semiquantitative information of gene expressions with very limited information on spatial patterns of their dynamics. In plant systems, however, methods have been developed to increase the spatial resolution of gene expression profiling data through isolation of specific cell types of given developmental fields (9, 10). However, the current challenge is to transform the static maps of gene expression into a unified and dynamic map (dynamic atlas) of gene activities, which can capture quantitative spatiotemporal information of all genes that are expressed in specific cell types. Given the fact that it is not possible to label and visualize more than three to four genes or proteins in a single organism, computational platforms are needed, which will facilitate comparisons of corresponding gene expression domains from a population of SAMs (Fig. 1K–P). However, the major challenge in obtaining spatiotemporal correspondence across SAMs is the inherent variability in SAM size, cell numbers, cell volumes, cell division rates, and orientations of cell divisions. Therefore, new computational thinking is required, first to map corresponding regions/gene expression patterns from several different SAMs into a single composite template and second to obtain temporal correspondences between different SAMs. This requires the development of global shape-matching algorithms and models for representing the dynamics of an entire class of SAMs.

2. Materials

1. Zeiss 310 or Zeiss 510 upright confocal microscope with multichannel imaging capability.
2. 63× achroplan water dipping objective lens (0.95 NA; Zeiss).
3. Clear plastic boxes (Part #: DG-0720; http://www.durphypkg.com/).
4. A pair of fine tweezers (#5 INOX; Dumont).
5. 1.5% Agarose.
6. Sterile water.
7. Transgenic plants with appropriate fluorescent constructs.
8. Fluorescent dyes, such as FM 1–43/FM4–64 (50 μg/ml; Molecular probes) series.

3. Methods

3.1. Live Cell Imaging

Live-cell imaging methods for monitoring SAM dynamics have been reviewed recently (11).

3.2. Image Processing and Data Analysis

To computationally analyze the patterns in individual cell growth and division, we need to segment out each individual cell from the cluster. After that, a cell tracking method is needed to get the correspondence between different optical cell slices at one time point as well as across two consecutive time points. In this section, we describe the segmentation methods followed by the cell tracking strategies that have been employed.

3.2.1. Image Registration

The time series of confocal Z-stacks of SAMs requires to be aligned and the 3D alignment can be done by using available software packages that utilize information theory to maximize mutual information across image stacks to register images at subpixel resolution (12). The 3D registration/alignment can also be done by using appropriate image registration module in commercial software packages, such as AMIRA (Visage Imaging). The registered stacks can then be reconstructed in 3Ds, rendered and animated to play continuous movies by using, for example, the Zeiss LSM3.2 software. The cells in the L1 layer, located at various depths on the curved surface, are projected onto a single 3D-reconstructed view by using maximum intensity projection.

3.2.2. Segmentation

Different methods have been employed to segment individual cells in the cluster which include Watershed Segmentation (13) and Level-Set-based segmentation (14).

Watershed Segmentation

Watershed transformation is often used to segment cell boundaries (13). Watershed treats the input image as a continuous field of basins (low-intensity pixel regions) and barriers (high-intensity pixel regions), and outputs the barriers that are the cell boundaries of all the cells in the image. The main problem of watershed algorithm is that it results in oversegmentation. So prior to applying the watershed algorithm, the input image undergoes H-minima transformation in which all the pixels below a certain threshold percentage "h" are discarded. The H-minima operator can be used to find a function, the set of all markers identifying attraction basins, the depth of which exceeds "h" so that we can eliminate local minima which are less than the threshold "h." In live imagery, the image quality degrades in the deeper layers of SAMs and in such cases, Watershed segmentation often tends to undersegment (merges number of neighboring regions/cells into one segmented region) the noisy regions in the images. Therefore, the Watershed segmentation threshold should be carefully chosen to produce better segmentation results.

Level-Set Segmentation

A level set is a collection of points over which a function takes on a constant value. In this case, the segmentation process starts with an initialized closed curve (known as an "Active Contour"), which then evolves towards the boundary of the cells until a local optimum is reached. At this point, every cell is segmented in the SAM structure and individual cells are represented by separate level-set functions. Different variations and improved versions of the original level-set method are available which can be used in the segmentation of SAM image slices depending on the noise content of the images and other factors (15, 16).

3.2.3. Plant Cell Lineage Computation: Tracking

In this section, a number of cell tracking techniques that are commonly used for cell lineage computation are described. The goal is to track each cell in the 4D space-time volume so as to obtain correspondences of cells across time (i.e., to provide a mapping between any two segmented cells in Fig. 2i). The method has to be robust to take into account new cells that arise due to cell division events and noisy images. The relative advantages and disadvantages of these methods are also briefly explained in the context of cell tracking in multilayered, multicellular tissues, such as shoot meristems.

Level-Set-Based Tracking

Level sets, as described in step 2 in Subheading 3.2.2, have been a very popular method for tracking cells through deformations and cell divisions (17–20). However, they are all limited to tracking cells in the 2D image plane rather than the 3D volume. There are several other reasons why the level-set-based tracking method cannot be successfully used in tracking SAM cells. First, the cells are in close contact with each other. Second, the SAM cells share similar features with respect to their shapes and sizes. Third, large parts of images may be noisy at a particular time instant due to a low signal-to-noise ratio.

Softassign Procrustes Algorithm

Given the sets of segmented cells from the time lapse imagery, tracking of cells over time is essentially a point matching problem. One of the most popular solutions to point matching problem has been the Softassign Procrustes algorithm, which has been applied to compute cell lineages, and it has been improved further to detect cell divisions (21, 22). The Softassign method uses the information on point location to simultaneously solve both the problem of global correspondence as well as the problem of affine transformation between two time instants iteratively. Although this method can be applied in aligning global features, it can produce errors in finding the local correspondences of individual cells.

Other Methods

The segmentation and tracking method using coupled active surfaces is described in ref. 23 which addresses the problem of tracking cells in 3D. Padfield et al. (24) approached the problem of cell tracking as a spatiotemporal segmentation task that can yield more accurate cell segmentation results than frame-by-frame processing. A combined mean shift-based tracking method was presented in ref. 25 that can establish migrating cell trajectories. But this method uses the phase-contrast video microscopy which is not effective for imaging individual cells in a multilayer cell cluster like the SAM. Incorporation of topological constraints (26, 27) in level-set-based tracking can effectively prohibit the cases of merging of cells and at the same time can allow cell division events, thereby producing more robust lineages. A recent method has used images acquired from multiple angles for tracking cell lineages (28). This imaging and image processing pipeline(MARS-ALT) could track around twelve hundred cells in a floral meristem over a seventy hour time period. The time gap between successive observations was twenty four hours.

Graph Matching-Based Tracking

Calculation of local neighborhood structures of cells is crucial in tracking cells of similar shapes that lie in close proximity as in the case of SAMs (Fig. 2iii). In this approach (29), information about the geometrical and topological interrelationships is utilized between local neighborhood structures of cells. Here, the heuristic graph matching method is described, wherein the matching problem is solved in a progressive manner (i.e., cell by cell) by obtaining correspondences from local graphs generated at different time instants. The local geometrical and topological features of cells are exploited to generate graphs of the local neighborhood of each cell (Fig. 2iii A, B). This process is followed by matching of the relative positional information of cells, such as the length and orientation of the edges with respect to their nearest neighbors to find the most similar local graph pair between two time instants. This process provides a seed (initial) cell pair (Fig. 2iii C, D) and the seed pair is used as a starting point to calculate similarities between local regions in the graph by progressively moving outwards (from this seed pair) to obtain correspondences of neighboring cells (Fig. 2iii E, F). This process is continued recursively to find correspondences of all cells. The method described above has greater robustness over the global matching methods (21), especially in cases where the images are noisy which in turn results in poor segmentation. Every cell in the image stack is usually represented in multiple consecutive optical slices obtained along Z-dimension which provides an opportunity to track the same cell several times. This method also utilizes integration or fusion of tracking results over the entire 4D image stack to improve the tracking efficiency, unlike earlier methods that have employed tracking on a single layer of cells or 3D-reconstructed surface layer (21). In fact, the integration or fusion of tracks from three consecutive slices greatly improves the tracking efficiency by accounting for cells that were not properly segmented in one of the three slices due to poor image quality. It is expected that the topology of a given local graph will change upon cell division. Therefore, the changes in topology of local graphs act as good indicators of cell division events and it has been used to detect cell division events.

3.2.4. Quantitative Growth Models at the Local/Cellular Level: Cell Volume Estimation

Measurement and quantification of the volume of individual cells of SAMs can be challenging. There are several methods of volume estimations for individual cells, such as impedance method (30) and light microscopy-based methods (31). Methods such as in ref. 32 are used to study changes in cell volumes in cell monolayers. But it is a much more challenging problem in the case of SAM, wherein cells are densely clustered. In such cases, the most popular practice is to use confocal laser scanning microscopy (CLSM) to image cell slices at a very high spatial resolution and then reconstruct the 3D volume of the cell from those serial optical slices. The method described in (28) accurately reconstructs the plant meristems.

This method uses a data set that consists of image slices acquired by imaging the tissue from multiple angles. These image slices are then computationally merged and segmented to obtain accurate 3D cellular reconstruction.

A recent method describes estimation of cellular volumes from confocal slices (33). This method assumes a prior ellipsoidal model (minimum volume enclosing ellipsoid, MVEE) for each individual cell in the SAM, and estimates the parameters of these ellipsoids and the resulting cell volumes by fitting segmented cell contours to them (33). Through computation of cell lineages and estimated cell volumes at each time instant, this method also provides the rate of change in cellular volumes. In fact, there can be several different ellipsoidal approximations for the segmented cells, such as MVEE, maximum volume-inscribed ellipsoids (MVIE), least-square ellipsoids, etc. Depending on the quality of the segmentation output, one approximation is preferred over the other. For example, least-square ellipsoids can be a better model for cell slices obtained from Watershed segmentation.

3.2.5. Quantification and Modeling of Global Growth Patterns

The global shape of the SAM and its deformation over time can provide significant insights into the growth processes at a far less computational cost compared to tracking every single cell quantitative representation of global growth dynamics is particularly important in studying the local control of global shape change. A recent study describes a method to reconstruct global shape of SAMs (34). This involves the following steps.

1. Extract boundary points from image.
2. Compute curvature values for each boundary point in vertical and horizontal directions.
3. Perform curvature covariance matrix analysis across the shape space.
4. Save detected regions of growth as separate shapes.
5. Compute the rate of change of each detected primordia across time.

The system is made up of two main modules: the detection module (DM) and the growth computation module (GCM). Given a stack of slices, the DM extracts boundary points from all the slices and also detects organ primordia. The output of the DM is a list of shape sequences. Each shape sequence in this list is made up of a set of shape vectors. A shape vector contains the set of points representing the coordinates of a SAM at a given depth. The GCM uses the shape vectors of a given SAM at consecutive time instants to compute the rate of shape change.

Acknowledgments

We thank the microscopy core facility of center for plant cell biology (CEPCEB) and institute of integrative genome biology (IIGB), University of California, Riverside. This work is funded by National Science Foundation grants (IOS-0820842 to GVR and IIS-0712253 to ARC).

References

1. Reddy GV (2008) Live-imaging stem-cell homeostasis in the Arabidopsis shoot apex. Curr Opin Plant Biol 11:88–93
2. Meyerowitz EM (2007) Genetic control review of cell division patterns in developing plants. Cell 88:299–308
3. Steeves TA, Sussex IM (1989) Patterns in plant development. Cambridge University Press, New York
4. Barton MK (2010) Twenty years on: the inner workings of the shoot apical meristem, developmental dynamo. Dev Biol 341:95–113
5. Reddy GV, Heisler MG, Ehrhardt DW, Meyerowitz EM (2004) Real-time lineage analysis reveals oriented cell divisions associated with morphogenesis at the shoot apex of *Arabidopsis thaliana*. Development 131:4225–4237
6. Grandjean O, Vernoux T, Laufs P, Belcram K, Mizukami Y, Traas J (2003) In Vivo Analysis of Cell Division. Cell Growth, and Differentiation at the Shoot Apical Meristem in Arabidopsis, Plant cell 16:74–87
7. Heisler MG, Ohno C, Das P, Sieber P, Reddy GV, Long JA, Meyerowitz EM (2005) Auxin transport dynamics and gene expression patterns during primordium development in the Arabidopsis Inflorescence Meristem. Curr Biol 15:1899–1911
8. Reddy GV, Meyerowitz EM (2005) Stem-cell homeostasis and growth dynamics can be uncoupled in the Arabidopsis shoot apex. Science 310:663–667
9. Yadav RK, Girke T, Pasala S, Xie M, Reddy GV (2009) Gene expression map of the Arabidopsis shoot apical meristem stem cell niche. Proc Natl Acad Sci USA 106:4941–4946
10. Brady SM et al (2007) A high-resolution root spatiotemporal map reveals dominant expression patterns. Science 318:801–806
11. Reddy GV, Roy-Chowdhury A (2009) Live-imaging and image processing of shoot apical meristems of *Arabidopsis thaliana*. Methods Mol Biol 553:305–316
12. Viola P, Wells WM III (1997) Alignment by maximization of mutual information. Int J Comput 24:137–154
13. Najman L, Schmitt M (1994) Watershed of a continuous function. Signal Process 38:99–112
14. Chan T, Vese L (2001) Active contours without edges. IEEE Trans Image Process 10:266–277
15. Nath S, Palaniappan K, Bunyak F (2006) Cell segmentation using coupled level sets and graph-vertex coloring. Lect Notes Comput Sci (MICCAI) 4190:101–108
16. Gelas A, Mosaliganti K, Gouaillard A, Souhait L, Noche R, Obholzer N, Megason SG (2009) Variational level-set with Gaussian shape model for cell segmentation. In: IEEE International Conference on Image Processing, p 1089–1092
17. Zimmer C, Labruyere E, Meas-Yedid V, Guillen N, Olivo-Marin J-C (2002) Segmentation and tracking of migrating cells in videomicroscopy with parametric active contours: A tool for cell-based drug testing. IEEE Trans Med Imaging 21:1212–1221
18. Dzyubachyk O, Niessen WJ, Meijering E (2007) A variational model for level-set based cell tracking in time-lapse fluorescence microscopy images. IEEE International Symposium on Biomedical Imaging: From Nano to Macro.
19. Ray N, Acton ST (2002) Active contours for cell tracking. In: Proceedings of the fifth IEEE southwest symposium on image analysis and interpretation, p 7–9
20. Mukherjee DP, Ray N, Acton ST (2001) Level set analysis for leukocyte detection and tracking. IEEE Trans Image Process 13:562–672
21. Gor V, Elowitz M, Bacarian T, Mjolsness E (2005) Tracking cell signals in fluorescent images. In: Proceedings of the IEEE Computer Society Conference on Computer Vision and Pattern Recognition
22. Rangarajan A, Chui H, Bookstein FL (2005) The Softassign procrustes matching algorithm. Information Process Med Imaging 29–42

23. Dufour A, Shinin V, Tajbakhsh S, Guillen-Aghion N, Olivo-Marin JC, Zimmer C (2005) Segmentation and tracking fluorescent cells in dynamic 3-D microscopy with coupled active surfaces. IEEE Trans Image Process 14:1396–1410
24. Padfield D, Rittscher J, Thomas N, Roysam B (2009) Spatiotemporal cell cycle phase analysis using level sets and fast marching methods. Medical Image Analysis 13(1):143–155
25. Debeir O, Van Ham P, Kiss R, Decaestecker C (2005) Tracking of migrating cells under phase-contrast video microscopy with combined mean-shift processes. IEEE Trans Med Imaging 24:697–711
26. Li K, Miller ED, Weiss LE, Campbell PG, Kanade T. (2006) Online tracking of migrating and proliferating cells imaged with phase-contrast microscopy. In: Proceedings of the 2006 Conference on Computer Vision and Pattern Recognition Workshop, p 65–72
27. Kanade T, Li K (2005) Tracking of migrating and proliferating cells in phase-contrast microscopy imagery for tissue engineering. In: Proceedings of the First International Workshop on Computer Vision for Biomedical Image Applications (CVBIA), p 24
28. Fernandez R, Das P, Mirabet V, Moscardi E, Traas J, Verdeil JL, Malandain G, Godin C (2010) Imaging plant growth in 4D: robust tissue reconstruction and lineaging at cell resolution. Nature Methods 7:547–553
29. Liu M, Yadav RK, Roy-Chowdhury A, Reddy GV. Automated tracking of stem cell lineages of arabidopsis shoot apex using local graph matching. Plant Journal Oxford, UK: Blackwell Publishing Ltd, 62(1):135–147
30. Nakahari T, Murakami M, Yoshida H, Miyamoto M, Sohma Y, Imai Y (1990) Decrease in rat submandibular acinar cell volume during ach stimulation. Am J Physiol 258:878–886
31. Farinas J, Kneen M, Moore M, Verkman A (1997) Plasma membrane water permeability of cultured cells and epithelia measured by light microscopy with spatial filtering. J Gen Physiol 110:283–296
32. Kawahara K, Onodera M, Fukuda Y (1994) A simple method for continuous measurement of cell height during a volume change in a single a6 cell. Jpn J Physiol 44:411–419
33. Chakraborty A, Liu M, Mkrtchyan K, Reddy GV, Roy-Chowdhury A (2010) Cell volume estimation from a sparse collection of noisy confocal image slices. In: Proceedings of the Seventh Indian Conference on Computer Vision, Graphics and Image Processing, p 183–189
34. Tataw O, Liu M, Yadav R, Reddy V, Roy-Chowdhury A (2010) Pattern analysis of stem cell growth dynamics in the shoot apex of Arabidopsis. In: IEEE International Conference on Image Processing, p 3617–3620

Index

Zhi-Yong Wang and Zhenbiao Yang (eds.), *Plant Signalling Networks: Methods and Protocols*, Methods in Molecular Biology, vol. 876, DOI 10.1007/978-1-61779-809-2,

MIX
Papier aus verantwortungsvollen Quellen
Paper from responsible sources
FSC® C105338

If you have any concerns about our products, you can contact us on
ProductSafety@springernature.com

In case Publisher is established outside the EU, the EU authorized representative is:
Springer Nature Customer Service Center GmbH
Europaplatz 3, 69115 Heidelberg, Germany

Printed by Libri Plureos GmbH
in Hamburg, Germany